Fungal Biology

About the Series

Fungal biology has an integral role to play in the development of the biotechnology and biomedical sectors. It has become a subject of increasing importance as new fungi and their associated biomolecules are identified. The interaction between fungi and their environment is central to many natural processes that occur in the biosphere. The hosts and habitats of these eukaryotic microorganisms are very diverse; fungi are present in every ecosystem on Earth. The fungal kingdom is equally diverse, consisting of seven different known phyla. Yet detailed knowledge is limited to relatively few species. The relationship between fungi and humans has been characterized by the juxtaposed viewpoints of fungi as infectious agents of much dread and their exploitation as highly versatile systems for a range of economically important biotechnological applications. Understanding the biology of different fungi in diverse ecosystems as well as their interactions with living and non-living is essential to underpin effective and innovative technological developments. This series will provide a detailed compendium of methods and information used to investigate different aspects of mycology, including fungal biology and biochemistry, genetics, phylogenetics, genomics, proteomics, molecular enzymology, and biotechnological applications in a manner that reflects the many recent developments of relevance to researchers and scientists investigating the Kingdom Fungi. Rapid screening techniques based on screening specific regions in the DNA of fungi have been used in species comparison and identification, and are now being extended across fungal phyla. The majorities of fungi are multicellular eukaryotic systems and therefore may be excellent model systems by which to answer fundamental biological questions. A greater understanding of the cell biology of these versatile eukaryotes will underpin efforts to engineer certain fungal species to provide novel cell factories for production of proteins for pharmaceutical applications. Renewed interest in all aspects of the biology and biotechnology of fungi may also enable the development of "one pot" microbial cell factories to meet consumer energy needs in the 21st century. To realize this potential and to truly understand the diversity and biology of these eukaryotes, continued development of scientific tools and techniques is essential. As a professional reference, this series will be very helpful to all people who work with fungi and should be useful both to academic institutions and research teams, as well as to teachers, and graduate and postgraduate students with its information on the continuous developments in fungal biology with the publication of each volume.

More information about this series at http://www.springer.com/series/11224

Bhim Pratap Singh • Garima Singh
Krishna Kumar • S. Chandra Nayak
N. Srinivasa
Editors

Management of Fungal Pathogens in Pulses

Current Status and Future Challenges

Springer

Editors
Bhim Pratap Singh
Department of Biotechnology
Mizoram University
Aizawl, Mizoram, India

Garima Singh
Department of Botany
College Veng
Pachhunga University College
Aizawl, Mizoram, India

Krishna Kumar
Division of Crop Protection
Indian Institute of Pulses Research
Kanpur, Uttar Pradesh, India

S. Chandra Nayak
Pearl Millet Pathology Laboratory
University of Mysore
Mysore, India

N. Srinivasa
Division of Plant Pathology
Indian Agricultural Research Institute
New Delhi, India

ISSN 2198-7777 ISSN 2198-7785 (electronic)
Fungal Biology
ISBN 978-3-030-35946-1 ISBN 978-3-030-35947-8 (eBook)
https://doi.org/10.1007/978-3-030-35947-8

This Springer imprint is published by the registered company Springer Nature Switzerland AG
The registered company address is: Gewerbestrasse 11, 6330 Cham, Switzerland

Preface

Pulses, due to their rich protein content, play an important role in maintaining the nutritional balance, and have become an integral part of versatile diets, including vegetarian diets, across the globe. The yield and quality of these crops are adversely affected due to various fungal pathogens, amounting around 100% yield losses in certain crops. Pulses are infected by approximately 100 fungal diseases all around the world. This book houses information on major fungal pathogens that cause significant losses, and on management strategies to reduce the incidence and severity of fungal diseases in pulse crops.

The present volume has 12 chapters dealing with major pathogens to pulses and their management. Chapter 1 deals with the role of plant growth-promoting rhizobacteria in the management of soil-borne fungal pathogens. Chapter 2 presents secondary metabolites which have also been proven to have antagonistic potential and are considered to have the ability to control fungal pathogens affecting pulses and other crops. Chapter 3 focuses on the management of fungal foliar diseases of arid legumes using integrated approach. Chapter 4 discusses omics approach to control *Fusarium* wilt of chickpea. Chapter 5 gives an overview of the management of fungal pathogens of chickpea, whereas Chapter 6 discusses the detection of wilt and root rot complex of important pulse crops with strategies to control them. Chapter 7 discusses the management strategies and diversity of *Phytophthora*, causing stem blight of pigeonpea. Chapter 8 reviews important foliar fungal diseases of pulses and their management strategies. Similarly, Chapter 9 talks about the role of soil and crop health management for cultivation of pigeon pea. Chapter 10 deals with the vital foliar diseases of chickpea with its science, epidemiology, and management practices. Chapter 11 focuses on the management of wilt in pigeonpea mainly caused by *Fusarium udum*. Lastly, Chapter 12 elaborates the use of biofertilizers as a sustainable tool for the management of fungal pathogens.

To sum it up, the present book volume gives comprehensive information about the prevalent fungal pathogens affecting pulses, and their management approaches for sustainable agriculture.

Mizoram, India — Bhim Pratap Singh
Mizoram, India — Garima Singh
Kanpur, India — Krishna Kumar
Mysore, India — S. Chandra Nayak
New Delhi, India — N. Srinivasa

Acknowledgments

Our sincere gratitude to the contributing authors who have contributed chapters in this book volume, *Management of Fungal Pathogens in Pulses: Current Status and Future Challenges*, on different aspects dealing with fungal pathogens of pulses and their management. We are also thankful to the series editors, Dr. Vijai Kumar Gupta and Prof. Maria G. Tuohy, for accepting our proposal and giving us a chance to bring out this volume. We are equally thankful to Springer Publishing for their continuous support and help received throughout from conceptualizing to production. The production team of Springer Nature is also acknowledged for their support and guidance. At last, we do agree that it is very much possible that some mistakes might be detected and we remain responsible for any mistake this volume may contain. Please feel free to inform us the same.

Mizoram, India — Bhim Pratap Singh
Mizoram, India — Garima Singh
Kanpur, India — Krishna Kumar
Mysore, India — S. Chandra Nayak
New Delhi, India — N. Srinivasa

Contents

Chapter 1
Plant Growth-Promoting Rhizobacteria in Management of Soil-Borne Fungal Pathogens

Parishmita Gogoi, Priyanka Kakoti, Juthika Saikia, Rupak K. Sarma, Archana Yadav, Bhim Pratap Singh, and Ratul Saikia

1.1 Introduction

Plant diseases reduce crop yield, product quality and contaminate food grains with toxic chemicals, causing a great economic loss (Zaidi et al. 2014). Soil-borne fungal pathogens cause root rot, leaf fall, wilting, etc. in plants and are responsible for the decline of yield in highly cultivated areas. These pathogens feed on organic soil residues which results in root rot, leading to death, and the growth rate of plants depends on their susceptibility to various environmental factors and hosts (Redman et al. 2001). *Pythium* spp., *Rhizoctonia* spp., *Fusarium* spp., *Sclerotinia sclerotiorum*, *Phoma* spp., and *Cylindrocarpon* spp. are few of the common pathogens of soil affecting most of the agricultural crops. The epidemiology of these pathogens is caused by a large number of physiochemical and biological factors. Most root rot-causing fungal pathogens can colonize and survive in soil (Pettitt et al. 1996). Development of a large number of fungicides has occurred due to numerous varieties and complexities of fungal diseases; unfortunately, resistance has already been developed by pathogens against these fungicides (Agrios 2005). The genetically resistant cultivar is another approach, but this is not feasible with time (Fry 2008).

Literature review for the last 50 years has shown that several microorganisms have grown competence against soil-borne pathogens and nematodes. PGPRs are studied and used in managing soil-borne fungal diseases in plants as they reduce

P. Gogoi · P. Kakoti · J. Saikia · R. K. Sarma · A. Yadav · R. Saikia (✉)
Biotechnology Group, Biological Science & Technology Division, CSIR-North East Institute of Science & Technology, Jorhat, Assam, India
e-mail: ratulsaikia@neist.res.in

B. P. Singh
Department of Biotechnology, Mizoram University, Aizawl, Mizoram, India

B. P. Singh et al. (eds.), *Management of Fungal Pathogens in Pulses*, Fungal Biology, https://doi.org/10.1007/978-3-030-35947-8_1

diseases by acting as biocontrol agents (Shaikh and Sayyed 2015). The PGPR stimulates other beneficial symbionts and protects plants in inhibiting the contaminated soils by degrading xenobiotics (Jacobsen 1997). Recently, researchers are working with beneficial microbe's potential for measuring of plant protection. The biocontrol agent use easy delivery, provides resistance mechanisms in the host, improves plant growth, and increases yield. These antagonists operate through parasitism, mycolytic enzymes, antibiosis, and competition for nutrients and space, secretion of volatile toxic metabolites, etc. Thus, PGPR biocontrol is recommended as a green approach; the commercial availability is very slow for a proportion of registration as biocontrol agents. Therefore, future research aims to develop genetically modified (GM) strains of PGPR to enhance plant growth-promoting activity and additional mechanisms for biocontrol (Glick and Bashan 1997; Blouin-Bankhead et al. 2004); it is necessary to understand the environmental factors that adequately act upon activity of PGPR and mechanism for biocontrol of some wild strains (Landa et al. 2004a, b; Berg and Smalla 2009) as this would be acting upon their inconsistent performance.

1.2 Biology of Soil-Borne Pathogen

Soil-borne pathogens survive as soil inhabitants (retain in the soil for comparably longer periods) and also as soil transients (retain in the soil for shorter periods). Soil-borne pathogens are survived in saprobes form. They are distributed in soil depending upon the history of cropping, production practices, and various other attributes. The root pathogen inoculum is present generally at the top 10 inches of the soil profile, in the vertical plane, whereas field inoculums are collected from a susceptible crop that grows in the horizontal plane. Soil type, pH, texture, moisture, temperature, and nutrient levels are some of the factors affecting the distribution of soil pathogens. Soils with poor irrigation facilities allow the growth of several soil-borne pathogens like *Phytophthora*, *Pythium*, and *Aphanomyces*. Similarly, *Fusarium* and *Verticillium* wilt also occur more frequently in damp soils rather than in dry soils (Deketelaere et al. 2017). *Streptomyces scabies* is one among the other pathogens occurring in wet soil. Some of the predominant soil-borne pathogens are cited in Table 1.1.

Table 1.1 Some predominant soil-borne pathogens

Fungi	Bacteria	Nematodes
Sclerotium rolfsii	*Erwinia*	*Meloidogyne*
Rhizoctonia solani	*Ralstonia*	*Heterodera*
Fusarium spp.	*Rhizomonas*	*Longidorus*
Pythium spp.	*Agrobacterium*	*Paratrichodorus*

1.3 Diseases Caused by Soil-Borne Pathogen

A diverse group of fungi and other organisms are the causal agents of soil-borne diseases. Genera *Pythium*, *Phytophthora*, *Rhizoctonia*, *Cylindrocladium*, and *Armillaria* are the most important which leads to root rots. The root rot diseases are distinguished by root system decay; some pathogens attack the juvenile roots, while others infect mature portions of the root system. Root rot symptoms include death of leaf, leaf fall, wilting, limb and branch death, and in extreme conditions full plant trends to die.

Root rot caused by *Rhizoctonia* is well-known as wire stem, damping-off, and crown or head rot. When the mature seedling is attacked by the fungus, the effect is less in the outer cortical tissues which produce elongate drab to the reddish-brown lesion. Infected area increases in length and width, spreading the disease to the whole plant causing death (Gonzalez et al. 2011).

Stem rots, head rots, and collar rot are incited by *Phytophthora*, *Fusarium*, *Rhizoctonia*, *Sclerotinia*, and occasionally *Aspergillus niger*, and the major symptom of these diseases is stem rot at ground level subsequently in death of leaves and the plant. *Fusarium oxysporum* and *Verticillium* spp. are the major fungi that cause wilts. Symptoms of wilt include internal necrosis of stems, vascular tissue and wilting of foliage. Similarly, bacterial pathogens also cause wilt disease in plants, resulting in loss of yield. Seedling blights and damping-off are caused by some fungal pathogens, *Phytophthora*, *Pythium*, *Sclerotium rolfsii*, *Fusarium*, and rarely *Rhizoctonia* spp. The fungi infect in different establishment stages of pre-emergence, post-emergence, or germination of the seedling. Damping-off disease by *Pythium* species like *P. debaryanum*, *P. graminicola*, *P. aphanidermatum*, and *P. ultimum* occurs in circular patterns as the fungi grow radically from the point of origin. *Phytophthora* damping-off disease, a low stem rot, is caused primarily by *P. fragaria*, *P. palmivora*, *P. cactorum*, and *P. syringae* where warmer soil temperatures (15–23 °C) are needed by the fungus for their rapid activity (Deadman 2017).

1.4 Management of Soil-Borne Disease

Management of soil-borne diseases require a thorough knowledge of host, pathogen, and environmental conditions. These three factors are responsible for the development of soil-borne diseases. The pathogens require viable inoculums to infect the host. The host needs to be exposed to the pathogen inoculums. For plant infection and pathogen growth, the environmental conditions should be suitable. These pathogen-host-environment dynamics help in constructing a disease management strategy (Shafique et al. 2016). For making a disease management strategy economical, potential crop loss, disease incidence assessments, and severity of diseases are key factors. It also needs regular and careful examination of symptomatic plants and fields. Disease management is also critical, e.g. the management of

Phytophthora root rot that requires early implementation of control measures. A management strategy in spite of being economically sound must also be safe, simple, and sufficiently effective to reduce diseases to acceptable levels. Management strategies of soil-borne diseases could be exclusion, eradication, and inoculum reduction. Use of resistant varieties, agronomic practices, chemical control, and biological control is useful for controlling this disease. Among those PGPR, ISR and systemic acquired resistance (SAR) are some of the important techniques (Beneduzi et al. 2012).

1.4.1 Soil-Borne Fungal Pathogens and PGPR

Substantial yield loss is caused by soil-borne the fungal pathogen (Oerke 2006). PGPR are environmental friendly management strategies (Weller et al. 2007). The usage of PGPR explicitly soil-borne fungal plant pathogen agents is a complementary strategy (Haas and Défago 2005; Weller 1998). Study shows that a wide range of PGPR protects against soil-borne fungal diseases (Saikia et al. 2003). The use of PGPR for their biocontrol effect in field conditions is often not steady enough which is one of its major limitations (Saikia et al. 2004a, b). Hence, some of the limitations of applying PGPR strains are sometimes not capable of surviving in their applied place or are not able to execute the specific biocontrol activity (Landa et al. 2001). One of the main reasons for their inconsistency that their survival rate is not the same in all types of ecosystems (Kravchenko et al. 1993; Picard and Bosco 2008; Berg and Smalla 2009). Biocontrol provided by PGPR involves competition, parasitism, antibiosis, etc. which comes under natural processes and is affected by abiotic and biotic factors (Weller et al. 2002, 2007; Haas and Défago 2005). The abiotic and biotic factors usually modify the interactions between plant, pathogen, and antagonist; thus, biocontrol agent efficiency is reduced on pathogens (Berg and Smalla 2009). Even if many abiotic soil factors influence the biocontrol mechanism (e.g. moisture, texture, pH, temperature, organic and inorganic constituents, etc.), there are very few experimental data of the interactions between antagonists and their soil-borne pathogens (Picard and Bosco 2008; Berg and Smalla 2009). Factors influencing the dynamics of populations in PGPR are not always affected by the biocontrol mechanisms governing PGPR efficacy. Pathogen suppression by PGPR occurs mainly by the activities involved in PGPR growth (Pathak et al. 2017).

Plant growth-promoting rhizobacteria enhance plant growth and development, also increases crop productivity. Rhizobacteria (PGPR) stimulate mechanisms that are broadly categorized as direct or indirect (Glick 1995). PGPR contributes directly to plant growth through phytohormone production like cytokines, gibberellins, and auxins, improving plant nutrition uptake by solubilizing minerals like iron and phosphorus, siderophore and enzyme productions, induction of systemic resistance, and lowering of ethylene level (Bhattacharyya and Jha 2012). The plant is indirectly benefited by PGPR as they enhance plant growth by controlling harmful microorganisms, including parasitism, antibiotic production, synthesis of extracellular enzymes

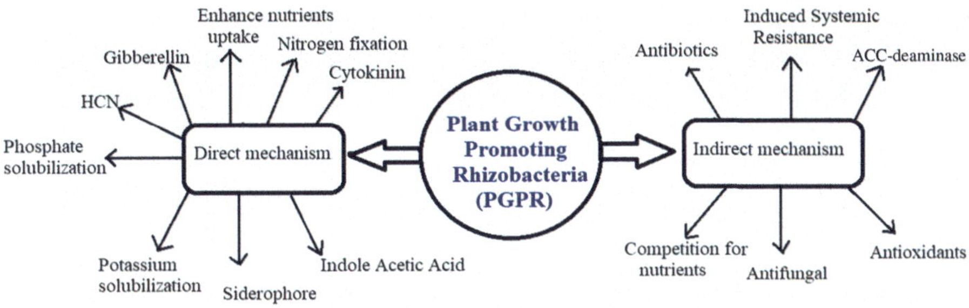

Fig. 1.1 Mechanism of plant growth promotion by rhizobacteria. Plant growth-promoting rhizobacteria (PGPR) promote plant growth directly by either assisting resource utilization of nitrogen, phosphorus, and other essential minerals or indirectly modulating plant hormone levels or by reducing the inhibitory effects of diverse pathogens on plant growth and development in the forms of biocontrol agents

for hydrolysing cell wall of fungi, decreasing pollutant toxicity, competition for nutrients and niches within the rhizosphere (Podile and Kishore 2006; Bhattacharyya and Jha 2012). The direct mechanism of PGPR includes the synthesis of plant hormones, nitrogen fixation, and phosphate solubilization (Ahemad and Kibret 2014). The indirect mechanism also includes biological controls, induced systemic resistance (ISR), antibiotics, competition for nutrients (Fig. 1.1).

Plant growth-promoting rhizobacteria involve various plant growth-promoting mechanisms and bacterial features that are important in facilitating the growth of the plant. It is controlled by 1-aminocyclopropane-1-carboxylate (ACC) deaminase enzyme as it cleaves ethylene precursor of plant, ACC into ammonia and α-ketobutyrate (Honma and Shimomura 1978). Plant ethylene level is decreased by ACC deaminase-producing organisms by lowering ACC level in plants, while ethylene present in maximum concentrations lead to growth inhibition or death in plants (Saikia et al. 2018).

In response to pathogen infections, plants produce an excess amount of ethylene in various stresses (Abeles et al. 1992). Symptoms shown by infection-causing pathogens which are seen in an infected plant appear as a direct result of pathogen imposing stress (Van Loon 1984). Increase in stress ethylene level of plants infected by pathogen generally results in damage to plants. Chemical inhibitors of ethylene synthesis decrease the severity of the infections, while severities of pathogen infections are increased by exogenous ethylene. Pretreating plants with ACC deaminase-producing rhizobacteria protects ethylene-caused damage in plants (Saikia et al. 2018).

1.4.2 Factors Influencing on Pathogen-PGPR Interactions

The factor of climate change specifically the increase in temperature has a link between PGPR and soil-borne pathogens and also on biocontrol efficacy interceded by PGPR (Table 1.2).

Table 1.2 Factors acting upon interactions of pathogen and PGPR

Sl. no.	Abiotic factors	Biotic factors
1	Soil physical and chemical characteristics	Target pathogen
2	Temperature	Host plant
3	Water availability	Insects
4	pH	Allelopathy
5	Moisture	Weeds
6	Quality and type of pesticides applied to the soil	Phytopathogens

1.4.3 Induced Resistance

Microorganisms are the environment-friendly approach used in controlling soil-borne diseases as biological control. The major approaches of biocontrol activity in PGPR are competition for antifungal metabolite production, nutrients, niche exclusion, and induced systemic resistance (ISR) (Lugtenberg et al. 2001). Plant growth-promoting rhizobacteria acting as biocontrol agents and its chief indirect mechanism. Antifungal metabolites produced by rhizobacteria are HCN, pyrrolnitrin, 2,4-diacetylphloroglucinol, phenazines, tensin, pyoluteorin, and viscosinamide (Bhattacharyya and Jha 2012). Rhizobacteria provide resistance against some pathogenic fungi, bacteria, and viruses interacting with plant roots (Lugtenberg et al. 2001).

Plant growth-promoting rhizobacteria trigger ISR in plants. Physical characteristics of ISR are similar to systemic acquired resistance (SAR). Plants activate their defence mechanisms against infection caused by a pathogenic agent, SAR (Pieterse et al. 2009). ISR is effective at managing diseases caused by various pathogens; it does not target specific pathogens (Saikia et al. 2005; Romera et al. 2019). ISR involves jasmonate and ethylene signalling pathways within the plant, and these hormones stimulate the host plant's defence responses to a range of pathogens (Verhagen et al. 2004). Other molecules, like O-antigenic side chain of the bacterial outer membrane protein lipopolysaccharide, cyclic lipopeptide surfactants, pyoverdine, chitin, flagellar proteins, β-glucans, and salicylic acid, have been summarized to act as signals for ISR. Van Peer et al. (1991) observed ISR in carnation plants protected systemically by *P. fluorescens* strain WCS417r against *F. oxysporum* f. sp. *dianthi.* ISR was also studied in cucumber plants (Wei et al. 1991). In cucumber leaves, rhizobacterial strains protect the leaves from anthracnose disease caused by *Colletotrichum orbiculare*. ISR mediated by rhizobacteria is similar to SAR induced by pathogens (Van Wees et al. 1997; Kannojia et al. 2019), involving viral, bacterial, and fungal pathogens, and also by insects and nematodes (Zehnder et al. 1997; Pozo and Azcon-Aguilar 2007; Bent 2006). It was also reported that in the same plant, the same strain provides resistance against several pathogens (Somers et al. 2004). The most studied rhizobacteria that trigger ISR are *Pseudomonas* and *Bacillus* spp. (Van Wees et al. 2008; Monaim 2012).

Systemic acquired resistance (SAR) and ISR are mediated by two different signalling pathways. ISR requires ethylene (ET)- and jasmonic acid (JA)-mediated signalling pathways, whereas SAR uses salicylic acid (SA) (Saikia et al. 2006). The signalling molecules accumulate and counter the defence responses (Ryals et al. 1996). ISR provides significantly lesser protection compared to SAR (Van Loon 2000). ISR depends on plant genotyping degree (Bloemberg and Lugtenberg 2001). According to Van Wees et al. (2000), SAR and ISR when used collaboratively provide better protection rather than acting alone upon pathogens. The utilization of exogenous SA also induces SAR in many plant species. Tissue necrosis is a common symptom for SAR activation (Vleesschauwer and Höfte 2009; Mishina and Zeier 2007). Pathogenesis-related (PR) proteins are specific sets of defence-related genes responsible for activating SA. Normally, ISR does not act upon the activation of PR genes. PR proteins are responsible for the enhanced defensive property of SAR (Van Loon 2007). The ethylene precursor, ACC, and also the methyl jasmonate (MeJA) provide pathogen resistivity (Shine et al. 2019). Different plant species studies have shown its ability to produce ISR in response to different PGPRs, and also the specific interaction among rhizobacteria and plants was studied (Van Loon 2007).

1.4.4 Other Control Methods

1.4.4.1 Cultural Method

Irrigation and fertilizer, when used together, improve the health of the plant. The use of ammonium bicarbonate, phosphatic fertilizer, phosphoric acid, and gypsum reduces the effect of soil-borne diseases in plants. The reduction of the disease requires good air circulation and good soil drainage within plants. Timely removal of dead or infected plants when disease occurs reduces inoculum build-up potential.

1.4.4.2 Crop Rotation

Soil-borne pathogens can exist in plant and soil debris for up to many years. Crop rotation can be applied to evade this problem as it helps in controlling the soil-borne inoculums. Pathogens are soil invaders that can help give the best result in crop rotation. However, crop rotation becomes less impractical when the pathogen resides in soil. In some causes of cropping systems, field tilting and field fallow are done for 6 months or a year (Veena et al. 2014).

1.4.4.3 Tillage Practices

Soil tilting can reduce the pathogen population by its burial or are dried in the exposed out layers. Deep ploughing is very useful in reducing the infection source. Before planting subsoiling is done to increase the yields of root rot-infected plants (Singh 2017).

1.4.4.4 Soil Amendments

Sawdust, straw, oil cake, etc. are organic amendments that are used effectively to manage diseases caused by *Aphanomyces*, *Pythium*, *Verticillium*, *Phymatotrichum*, *Macrophomina*, and *Phytophthora*. Useful microorganisms multiply in soil and help to suppress pathogens. Lime usage increases soil pH to 8.5 which reduces cabbage clubroot. Castor cake and neem leaves play a crucial role in reducing the foot rot of wheat.

1.4.4.5 Soil Solarization

Soil solarization is rise of soil temperature by sunlight. Various soil-borne pathogens like bacteria, fungi, and nematodes reduce the potential and inoculum for disease by inactivating near the soil surface due to soil solarization. *Verticillium* and *Fusarium* wilts are controlled by soil solarisation (Veena et al. 2014).

1.4.4.6 Chemical Control

The application of chemical fungicides is done to defend the plant from disease or eliminate a pathogen infecting the plant. Chemical control includes soil treatments, disinfestations of warehouses, cleaning of equipment, etc. Application of fungicides is in the form of liquid drenches, granules, or dust to the soil to eradicate diseases. They are applied in the fields through the irrigation system available. Nematodes are treated by chemical controls and volatile substances. Chemical fungicides mainly act as toxic barriers between host and pathogens. They are used as soil drenching, seed treatment, and soil fumigation. Propamocarb, prothiocarb, and metalaxyl are some of the frequently used fungicides. Chemical fungicides cause a lot of harm to the soil and plant along with reduction of diseases (Mahmood et al. 2016).

1.4.4.7 Resistance of Host Plant

Making a resistant plant is the most cost-effective and adequate method. It reduces the loss of yield, and also it reduces pollution and cuts off the disease controlling effort. Monogenic (vertical) resistance is a gene- or race-specific resistance that is capable of controlling only a few pathogens. On the other hand, polygenic (horizontal) resistance is a quantitative or non-specific resistance. It lasts longer and is not so adequate. Host resistance is useful when used together with chemical and cultural methods.

In transgenic approaches, genes are transformed for tolerating detrimental abiotic and biotic conditions, and for genes encoding enzymes like glucanases and chitinases acting upon fungi, viruses, and bacteria by using DNA technology. Various PR proteins, glucanases, and chitinase-coding genes are cloned, isolated, and expressed in plants; thus, the development of pathogens is resisted along with plant resistance.

1.4.4.8 Aerial Photography

It identifies objects in a higher range of land. This technique was first used by Colwell (1956). He used infrared aerial photography to identify rusts, citrus diseases, and small grain viral diseases. Panchromatic, normal, and infrared colour are the major films used in aerial photography. Ektachrome Aero Infrared (camouflage detection film) can portray the difference between the healthy and diseased colour patches in plants (Veena et al. 2014).

1.5 Conclusion

Management of soil-borne diseases can be successful and cost-effective if we have a detailed information/knowledge regarding crop, disease history, resistant levels, and environmental conditions. The increasing concern about nature and understanding the adverse effect of chemical use in the environment, non-chemical methods have been developed for the prevention of soil-borne diseases. PGPR offers an attractive alternative for sustainable approaches to agriculture. Credentials of diverse mechanisms involved in plant-rhizosphere microorganism interactions opened new possibilities to design strategies for improving crop yields. Subsequently, microbial strains that have plant growth-promoting traits are improved with the use of a biotechnological approach to create transgenic strains with multiple mechanisms of action. Comprehensive knowledge of plant-microbe interactions in the rhizosphere is necessary before utilizing PGPR as biofertilizers which establish a sustainable promotion of plant growth. Genes providing resistance to common and widely occurring soil-borne fungal pathogens normally lack economic importance in most cultivated plants. Alternatively, a strategy is evolved in plants that stimulate and support specific antagonistic microorganisms groups from lots of deleterious, beneficial, and neutral species in the environment of the rhizosphere. Thus, PGPRs are the most important antagonistic microorganisms selected since they are rich in nutrients released from plant roots, and they provide the first line of defence against soil-borne diseases (Weller et al. 2007; Cook et al. 1995). Identifying environmental factors influencing the disease management capability of these PGPRs would cater a base for enhanced alliance treatments of biocontrol with different control practices that are environmental friendly, both under climate scenarios of the present and future, making the farmers capable of managing soil-borne diseases and reducing the use of chemical pesticides.

Acknowledgement The work is supported by a Network Project, MLP-1005, sponsored by the Council of Scientific and Industrial Research, Ministry of Science and Technology, Government of India, New Delhi. The authors are also thankful to the Director of CSIR-NEIST, Jorhat, Assam, for providing necessary facilities to carry out the work and DBT-BIF Centre, CSIR-NEIST, Jorhat, for providing the computational facilities.

References

Abeles FB, Morgan PW, Saltveit ME Jr. Ethylene in plant biology. 2nd ed. New York: Academic Press. 1992.

Agrios GN. Department of plant pathology. Amsterdam: University of Florida, Elsevier Academic. 2005; p. 635.

Ahemad M, Kibret M. Mechanisms and applications of plant growth promoting rhizobacteria: Current perspective. J King Saud Unv. 2014;26:1–20.

Beneduzi A, Ambrosini A, Passaglia L. Plant growth-promoting rhizobacteria (PGPR): their potential as antagonists and biocontrol agents. Genet Mol Biol. 2012;35:1415–4757.

Bent E. Induced systemic resistance mediated by plant growth-promoting rhizobacteria (PGPR) and fungi (PGPF). In: Tuzun S and Bent E (eds) Multigenic and Induced Systemic Resistance in Plants. Springer Science, New York. 2006; p. 225–59.

Berg G, Smalla K. Plant species and soil type cooperatively shape the structure and function of microbial communities in the rhizosphere. Microbiol Ecol. 2009;68:1–13.

Bhattacharyya P, Jha D. Plant growth-promoting rhizobacteria (PGPR): emergence in agriculture. J Microbiol Biotechnol. 2012;1–24.

Bloemberg GV, Lugtenberg, BJ. Molecular basis of plant growth promotion and biocontrol by rhizobacteria. Curr Opin Plant Biol. 2001;4:343–50.

Blouin-Bankhead S, Landa BB, Lutton E, Weller DM, Mcspadden B. Minimal changes in rhizobacterial population structure following root colonization by wild type and transgenic biocontrol strains. Microbiol Ecol. 2004;49:307–318.

Colwell RN. Determining the prevalence of certain cereal crop diseases by means of aerial photography. Hilgardia. 1956;26:223–86.

Cook RJ, Thomashow LS, Weller DM, Fujimoto D, Mazzola M, Bangera G, Kim D-S. Molecular mechanisms of defense by rhizobacteria against root disease. Proc Natl Acad Sci, USA. 1995;92:4197–201.

Deadman M. Pythium and phytophthora damping-off and root rot. American Phytopathological Society. 2017; p. 48–50.

Deketelaere S, Tyvaert L, Franca SC, Hofte M. Desirable Traits of a Good Biocontrol Agent against Verticillium Wilt. Front Microbiol. 2017;8:1186.

Fry W. Phytophthora infestans: the plant and R gene destroyer. Mol Plant Pathol. 2008;9:385–402.

Gonzalez M, Pujol M, Metraux JP, Vicente GG, Bolton MD Orlando BH. Tobacco leaf spot and root rot caused by Rhizoctonia solani Kühn. Mol Plant Pathol. 2011;12:209–16.

Glick BR. The enhancement of plant growth promotion by free living bacteria. Can J Microbiol. 1995;41:109–17.

Glick BR, Bashan Y. Genetic manipulation of plant growth promoting bacteria to enhance biocontrol of phytopathogens. Biotechnol Adv. 1997;15:353–78.

Haas D, Défago G. Biological control of soil-borne pathogens by fluorescent pseudomonads. Nat Rev Microbiol. 2005;3:307–19.

Honma M, Shimomura T. Metabolism of 1-aminocyclopropane-1-carboxylic acid. Agri Biol Chem. 1978;43:1825–31.

Jacobsen CS. Plant protection and rhizosphere colonization of barley by seed inoculated herbicide degrading Burkholderia (Pseudomonas) cepacia DBO1 (pRO101) in 2,4-D contaminated soil. Plant Soil. 1997;189:139–44.

Kannojia P, Choudhary K.K, Srivastava A.K, Singh A.K. PGPR Bioelicitors: Induced Systemic Resistance (ISR) and Proteomic Perspective on Biocontrol. PGPR Amelioration in Sustainable Agriculture-Food Security and Environmental Management, Woodhead Publishing. 2019;67–84. https://doi.org/10.1016/B978-0-12-815879-1.00004-5

Kravchenko LV, Azarova TS, Dostanko OY. Effect of root exometabolites of wheat with different genome ploidy on growth of Azospirillum brasilense. Microbiol. 1993;62:517–20.

Landa BB, Navas-Cortés JA, Hervás A, Jiménez-Díaz RM. Influence of temperature and inoculum density of *Fusarium oxysporum* f. sp. *ciceris* on suppression of *Fusarium* wilt of chickpea by rhizosphere bacteria. Phytopathology. 2001;91:807–16.

Landa BB, Navas-Cortés JA, Jiménez-Díaz RM. Influence oof temperature on plant rhizobacteria interactions related to biocontrol potential for suppression of *Fusarium* wilt of chickpea. Plant Pathol. 2004a;53:341–52.

Landa BB, Navas-Cortés JA, Jiménez-Díaz RM. Integrated management of *Fusarium* wilt 1010 of chickpea with sowing date, host resistance, and biological control. Phytopathology. 2004b;94:946–60.

Lugtenberg BJJ, Dekkers L, Bloemberg GV. Molecular determinants of rhizosphere colonization by Pseudomonas. Annu Rev Phytopathol. 2001;39:461–90.

Mahmood I, Imadi SR, Shazadi K, Gul A. Effects of pesticides on environment. In: Hakeem KR, et al (ed). Plant, soil and microbes; Springer International Publishing Switzerland. 2016;253–69. https://doi.org/10.1007/978-3-319-27455-3_13

Mishina TE, Zeier J. Pathogen-associated molecular pattern recognition rather than development of tissue necrosis contributes to bacterial induction of systemic acquired resistance in Arabidopsis. Plant J. 2007;50:500–13.

Monaim MFA. Induced systemic resistance in tomato plants against *Fusarium* wilt disease. Int Res J Microbiol. 2012;3(1):14–23.

Oerke EC. Crop losses to pests. J Agric Sci. 2006;144:31–43.

Pathak R, Shrestha A, Lamichhane J, Gauchan DP. PGPR in biocontrol: mechanisms and roles in disease suppression. Int J Agron Agri R. 2017;11:69–80.

Pettitt TR, Parry DW, Polley RW. Effect of temperature on the incidence of nodal foot rot symptoms in winter wheat crops in England and Wales caused by *Fusarium culmorum* and Microdochium nivale. Agric For Meteorol. 1996;233–42.

Picard C, Bosco M. Genotypic and phenotypic diversity in populations of plant-probiotic *Pseudomonas* spp. colonizing roots. Naturwissenschaften. 2008;95:1–16.

Pieterse CMJ, Leon-Reyes A, van der Ent S, van Wees SCM. Networking by small-molecule hormones in plant immunity. Natur Chem Biol. 2009;5:308–16.

Podile AR, Kishore GK. Plant growth promoting rhizobacteria. In: Gnanamanickam SS (ed). Plant-associated bacteria, Springer, The Netherlands. 2006;195–230. https://doi.org/10.1007/978-1-4020-4538-7_6

Pozo MJ, Azcón-Aguilar C. Unraveling mycorrhiza- induced resistance. Curr Opin Plant Biol. 2007;10:393–8.

Redman RS, Dunigan DD, Rodriguez RJ. Fungal symbiosis: from mutualism to parasitism, who controls the outcome, host or invader. New Phyto. 2001;705–16.

Romera FJ, García MJ, Lucena C, Medina AM, Aparicio MA, Ramos J, Alcántara E, Angulo M, Vicente RP. Induced systemic resistance (ISR) and fe deficiency responses in dicot plants. Front Plant Sci. 2019;287:1–17.

Ryals JA, Neuenschwander UH, Willits MG, Molina A, Steiner H-Y, Hunt MD. Systemic acquired resistance. Plant Cell. 1996;8:1808–19.

Saikia J, Sarma RK, Dhandia R, Yadav A, Gupta VK, Bharali R, Saikia R. Alleviation of drought stress in pulse crops with ACC deaminase producing rhizobacteria isolated from acidic soil of Northeast India. Scientific Reports. 2018;8:3560.

Saikia R, Kumar R, Arora DK, Gogoi DK, Azad P. *Pseudomonas aeruginosa* inducing rice resistance *Rhizoctonia solani*: production of salicylic acid and peroxidases. Folia Microbiol. 2006;51(5):375–80.

Saikia R, Kumar R, Singh T, Srivastava AK, Arora DK, Lee MW. Induction of Defense Related Enzymes and Pathogenesis Related Proteins in *Pseudomonas fluorescens*-Treated Chickpea in Response to Infection by *Fusarium oxysporum ciceri*. Mycobiol. 2004a;32:47–52.

Saikia R, Singh K, Arora DK. Suppression of *Fusarium*-wilt and charcoal rot of chickpea by *Pseudomonas aeruginosa*. Indian J Microbiol. 2004b;44:10–14.

Saikia R, Singh T, Kumar R, Srivastava J, Srivastava AK, Arora DK. Role of salicylic acid in systemic resistance induced by *Pseudomonas fluorescens* against *Fusarium oxysoporum*. sp. *ciceri* in chickpea. Microbiol Res. 2003;158:203–13.

Saikia R, Srivastava AK, Singh K, Arora DK. Effect of Iron availability on induction systemic resistance to *Fusarium* wilt of chickpea by *Pseudomonas* spp. Mycobiology. 2005;33:35–40.

Shafique HA, Sultana V, Haque SE, Athar M. Management of soil-borne diseases of organic vegetables. J plant protection research. 2016;56:221–30.

Shaikh SS, Sayyed RZ. Role of plant growth-promoting rhizobacteria and their formulation in biocontrol of plant diseases. Springer. 2015;337–51.

Shine MB, Xiao X, Kachroo P, Kachroo A. Signaling mechanisms underlying systemic acquired resistance to microbial pathogens. Plnt Sci. 2019;279:81–6.

Singh RS. Disease management – The practices. Introduction to principles of plant pathology. Oxford & IBH Publishing Co. Pvt. Ltd. New Delhi. 2017.

Somers E, Vanderleyden J, Srinivasan M. Rhizosphere bacterial signalling, a love parade beneath our feet. Crit Rev Microbiol. 2004;30:205–35.

Van Loon LC. Regulation of pathogenesis and symptom expression in diseased plants by ethylene. Ethylene: Biochemical, Physiological and Applied Aspects, An International Symposium, Oiryat Anavim, Israel. 1984; p.171–80.

Van Loon LC, Bakker PAHM, Pieterse CMJ. Systemic resistance induced by rhizosphere bacteria. Annu Rev Phytopathol. 1998;36:453–83.

Van Loon LC. Systemic induced resistance. In: Slusarenko AJ, Fraser RSS and Van Loon LC (eds) Mechanisms of resistance to plant diseases. Kluwer Academic Publishers, Dordrecht. 2000; p.521–74.

Van Loon LC. Plant responses to plant growth-promoting rhizobacteria. Eur J Plant Pathol. 2007;119:243–54.

Van Peer R, Niemann GJ, Schippers B. Induced resistance and phytoalexin accumulation in biological control of *Fusarium* wilt of carnation by Pseudomonas sp. strain WCS417r. Phytopathology. 1991;91:728–34.

Van Wees SCM, De Swart EAM, Van Pelt JA, Van Loon LC, Pieterse CMJ. Enhancement of induced disease resistance by simultaneous activation of salicylate- and jasmonate-dependent defense pathways in *Arabidopsis thaliana*. Proc Natl Acad Sci, USA. 2000;97:8711–6.

Van Wees SCM, Pieterse CMJ, Trijssenaar A, Van't Westend YAM, Hartog F, Van Loon LC. Differential induction of systemic resistance in Arabidopsis by biocontrol bacteria. Mol Plant Microbe Interact. 1997;10:716–24.

Van Wees SCM, Van der Ent S, Pieterse CMJ. Plant immune responses triggered by beneficial microbes. Curr Opin Plant Biol. 2008;11:443–8.

Veena DR, Priya HR, Khatib RM, Joythi D. Soilborne diseases in Crop Plants and their Management. J Agri and Allied Sc. 2014;2319–9857.

Verhagen BWM, Glazebrook J, Zhu T, Chang HS, van Loon LC, Pieterse CMJ. The transcriptome of rhizobacteria-induced systemic resistance in Arabidopsis. Mol Plant-Microbe Interac. 2004;17:895–908.

Vleesschauwer D, Höfte M. Rhizobacteria-induced systemic resistance. Adv Bot Res. 2009;51:223–81.

Wei G, Kloepper JW, Tuzun S. Induction of systemic resistance of cucumber to Colletotrichum orbiculare by select strains of plant growth-promoting rhizobacteria. Phytopathology. 1991;81:1508–12.

Weller DM, Landa BB, Mavrodi OV, Schroeder KL, De La Fuente L, Blouin-Bankhead S, Allende-Molar R, Bonsall RF, Mavrodi DM, Thomashow LS. Role of 2,4-diacetylph-loroglucinol-producing fluorescent Pseudomonas spp. in plant defense. Plant Biol. 2007;9:4–20.
Weller DM, Raaijmakers JM, McSpadden Gardener BB, Thomashow LS. Microbial populations responsible for specific soil suppressiveness to plant pathogens. Annu Rev Phytopathol. 2002;40:309–48.
Weller DM. Biological control of soilborne plant pathogens in the rhizosphere with bacteria. Annu Rev Phytopathol. 1998;26:379–407.
Zaidi A, Ahmad E, Khan S. Role of Phosphate-Solubilizing Microbes in the Management of Plant Diseases. Springer. 2014;225–56.
Zehnder G, Kloepper J, Yao C, Wei G. Induction of systemic resistance in cucumber against cucumber beetles (Coleoptera, Chrysomelidae) by plant growth-promoting rhizobacteria. J Econ Entomol. 1997;90:391–6.

Chapter 2
Exploration of Secondary Metabolites for Management of Chickpea Diseases

Deepika Sharma, Sachin Gupta, Moni Gupta, and Baby Summuna

2.1 Introduction

Pulses are important components of the farming system both ecologically and nutritionally (human and animal). Although pulse crops are more important due to its nutritional value, there has not been any remarkable increase in area under its cultivation and production during 1950–2010. However, a significant increase in area under pulse crop cultivation and production has been recorded from 2010 to 2011 onward. The production of pulse crops has increased by approximately 68% at 764 kg/ha during the year 2014 from 441 kg/ha during 1951. Over a dozen pulse crops are grown annually all over the country in about 22–23 million hectares of area, producing 13–15 million tons of pulses. However, the prices of pulses have skyrocketed over the last few years making life difficult for the poor peoples to afford. One of the important reasons behind the price rise has been the fact that over the years, the production of pulses has declined due to the attack of diseases and insects. Around 8–10% of pulse crops are lost every year due to ravages of diseases alone costing nearly 1000 crores to the National Exchequer. The reduction of losses caused by diseases is, therefore, an important component of crop production technology.

D. Sharma · M. Gupta
Division of Biochemistry, Sher-e-Kashmir University of Agricultural Sciences and Technology of Jammu, Jammu and Kashmir, India

S. Gupta (✉)
Division of Plant Pathology, Sher-e-Kashmir University of Agricultural Sciences and Technology of Jammu, Jammu and Kashmir, India

B. Summuna
Department of Plant Pathology, Faculty of Agriculture-Wadura, Sher-e-Kashmir University of Agricultural Sciences and Technology of Kashmir, Jammu and Kashmir, India

B. P. Singh et al. (eds.), *Management of Fungal Pathogens in Pulses*, Fungal Biology, https://doi.org/10.1007/978-3-030-35947-8_2

Among pulses, chickpea (*Cicer arietinum* L.) is the world's fourth significant pulse crop after peas, common bean, and soybean. Chickpea is a rich supplement to the cereals in developing countries due to its high nutritional value. Chickpea is considered to be important because of the high level of protein content present in it, i.e., about 40% of its weight. Moreover, chickpea crop has various health benefits such as lessening the danger of cardiovascular diseases, cancer, diabetes, and other health problems. Chickpea alone contributes the largest share of ≈85.64% and 84.87% in India's export market of pulses during the year 2014–2015 and 2015–2016, respectively. This crop is mainly grown for its edible seeds which are high in protein content and forage production (Yadav et al. 2011). India contributes 75% of the world's total production of chickpea (Mahajan et al. 2018), and the crop accounts for 48% of the total pulse production in India (Anonymous 2015).

Chickpea's productivity remained stagnant from the last few decades due to the susceptibility of cultivars to various soilborne diseases and insects. In temperate regions, yield losses due to insects and diseases range from 5% to 10%, whereas in tropical regions, it is 50% to 100% (Van Emden et al. 1988). In this context, disease management in cereals and pulse crops is very important to alleviate the problem of shortages of food to feed the ever-growing population and to improve food production efficiency. Many microbial pathogens including airborne and soilborne pathogens have been reported to affect the chickpea crop. In chickpea, the development of resistance against the soilborne fungal pathogens is the major research efforts that have been made as compared with the foliar fungal pathogens. Foliar pathogen management gained the least important because they don't cause much yield loss. The list of common fungal diseases of chickpea is summarized in Table 2.1.

2.2 Role of Endophytic Bacteria in the Management of Plant Diseases

Endophytes are microorganisms, both bacteria and fungi, that reside within the plant host tissues without causing any harm to the host (Hallmann et al. 1997). Many endophytic bacteria are being used as promising biocontrol agents against the plant pathogens (Passari et al. 2015a, 2016, 2017). Endophytic bacteria colonize in the internal tissues of the host plant for improving crop health and its protection (Pavlo et al. 2011). Endophytic bacteria can promote the growth of the plant by altering its physiology which includes osmotic regulation, increased uptake of certain minerals, changes in stomatal responses, and nitrogen accumulation and metabolism (Compant et al. 2005). AitBarka et al. (2002) reported that endophytic bacteria trigger induced systemic resistance (ISR)-based plant growth promotion.

As have been reported by Pleban et al. (1995), *P. fluorescens* and *Bacillus* sp. effectively inhibited the growth of *Rhizoctonia solani* (46–56%, in bean), *Pythium ultimum*, and *Sclerotium rolfsii* (26–79%) plant pathogens. Experiments that have been conducted on various crops such as oilseed rape (Alstrom 2000), tomato (Chen et al. 1995), cotton (Liu et al. 1995), cucumber (Safiyazov et al. 1995), and

Table 2.1 Common fungal diseases of chickpea with their host and expected symptoms

S. No	Disease	Causal organisms	Host	Symptoms
1.	*Ascochyta* blight	*Didymella rabiei (anamorph: Ascochyta rabiei)*	Chickpea	First gray circular spots appear on leaves and pod that later turn dark brown with black borders. Black dots (pycnidia) are also present in advanced lesions in concentric rings
2.	*Fusarium* wilt	*Fusarium oxysporum* f. sp. *ciceris*	Chickpea	Leaves droop and appear pale and plants collapse and lie flat on the ground. Often, brown discoloration of internal root tissue is visible when the root splits into two vertically from the collar region
3.	Powdery mildew	*Leveillula taurica*	Chickpea	Small patches of white powdery coating initially develop on both surfaces of the older leaves. Affected leaves turn purple and then die. Stems, young leaves, and pods are also covered with the powdery coating The plant may also lose leaves too early in the season and produce seeds that are smaller than normal
4.	White mold stem and crown rot of chickpea	*Sclerotinia (S. minor, S. sclerotiorum, and S. trifoliorum)*	Chickpea	Visible white mycelium grows around the stem on the soil surface Black bodies (sclerotia) appear in various shapes and sizes on dead or dying chickpea stems Infected stems become pale in color, like bleaching, and the symptoms spread both upward and downward along the stems
5.	Black root rot	*Fusarium solani*	Chickpea	General symptoms include yellowing and wilting of scattered plants, rotten root system, shedding of finer roots, and remaining roots turning black
6.	*Pythium* seedling and root rot	*Pythium* sp.	Chickpea	Root tissue may die and become discolored, leading to less branching and fewer feeder roots. Low emergence and seed rot could occur. Discoloration of the crown and hypocotyl's tissue may be observed as rotting progresses. Stunting of plants is common and some plants can die before flowering, leading to reduced yield

(continued)

Table 2.1 (continued)

S. No	Disease	Causal organisms	Host	Symptoms
7.	Downy mildew	*Peronospora* sp.	Chickpea	The disease is often exhibited in a few branches, leading to curled or twisted leaves and dwarfed tips The symptoms may appear on any aerial part of the plant with white mycelial patches appearing first on the lower leaf surfaces and then chlorotic to yellow spots on the upper surface Fine, dirty, pinkish tufts of fungal growth are often formed on leaf surfaces under cool and humid conditions, which may disappear when dry conditions take over, resulting in yellowing symptoms The chlorotic spots then become dark and brittle Stunting and bushy apical growth with small leaflets is typical The affected plants can also lose all their leaves, resulting in reduced yield and seed size
8.	Gray mold(*Botrytis* stem and pod rot)	*Botrytis cinerea*	Chickpea	Water-soaked lesions on any aerial parts of the plant are indicative of infection, with the growing tips and flowers being the most susceptible After some time, the lesions change to gray or dark brown and take on a fuzzy appearance as a result of the hairy sporophores and masses of conidia The stem may be girdled by the lesions and the leaves often turn into a rotting mass. The dead tissues could have tiny, black sclerotia that form on them If the disease moves to pods, the seed may not form or they may shrivel or become discolored Frequently flowers drop and the pod formation could be unfavorable, leading to low grain yields
9.	Rust	*Uromyces ciceris-arietini* hellow	Chickpea	At first, small, round, brown spots (pustules) appear The pustules are sometimes surrounded by chlorotic halos. They often appear in a ring pattern. These may combine later and turn dark If the infection is severe, the leaves may drop off

peas (Sturz et al. 1999) by using endophytes such as *Pseudomonas* and *Bacillus* spp. against the fungal pathogens provide evidence of plant growth protection and promotion by the introduced endophytic bacteria.

2.3 Secondary Metabolite Production

Interspecies interactions in nature are often exhibited by microorganisms. Competition for space and nutrients results in interspecies interaction prompting the generation of secondary metabolites for improving their growth and development (Passari et al. 2019; Calvo et al. 2002). Competition among microbes for space and resources serves to be the major driving force for secondary metabolite production (Oh et al. 2005). Studies on secondary metabolite production by microbes and their application in suppressing plant diseases are gaining much significance in farming systems (Gohain et al. 2019). Because of the increased concerns on environmental pollution, pathogen resistance, and high plant security costs, secondary metabolites produced by microbes have been developed as commercial pesticides and can be used as an alternative to chemical fungicides. These metabolite products can also be utilized as bactericides, fungicides, and insecticides (Singh et al. 2019).

2.3.1 Secondary Metabolites Associated with Rhizobacteria in the Management of Plant Diseases

Biocontrol using microbial antagonists is becoming a critically needed component of plant disease management, particularly in reducing the risk of soilborne diseases using potential microorganisms (Mishra et al. 2016; Nautiyal 2000; Meki et al. 2009). At present, control of soilborne and seed-borne pathogens has been achieved mainly through the use of bacterial and fungal antagonists. Some rhizobacteria especially *Pseudomonas* spp. and *Bacillus* spp. from the plant rhizosphere are effective against the plant pathogens and also help the plants to acquire nutrients (Gopalakrishnan et al. 2011). Moreover, the use of biological control agents is much safer for the environment than synthetic or chemical pesticides.

Various mechanisms have been involved in antagonism, like cell wall-degrading enzymes (pectolytic enzymes, cellulases, xylanases, and glycosidic hydrolases) and siderophores that cannot only bind iron but also contribute to suppression of diseases of the plant (Passari et al. 2015b; Deshwal et al. 2003). Kravchenko et al. (2002) suggested that siderophores produced by microbes may also enhance plant growth by competitively inhibiting iron uptake system by fungal pathogens. Biological control agents also produce different types of volatile and diffusible antifungal metabolites which have the potential to suppress diseases caused by a fungal pathogen in various pathosystems (Yang et al. 2009). *Trichoderma* sp. has greater potential to control chickpea wilt under field as well as in polyhouse conditions, but its efficacy is not almost the same everywhere (Kaur and Mukhopadhayay 1992).

Rhizobacteria are ideal biocontrol agents that reside in the rhizosphere that give frontline protection to the roots against the pathogen entry. Rhizobacteria have received special attention as they are excellent root colonizers and have the potential to induce plant's defense mechanism through the production of various pathogenesis-related (PR) proteins (Kumar et al. 2010). *Bacillus* spp., a gram-positive rhizobacteria, are potential biocontrol agents because of its abundance in the rhizosphere and have the potential to produce active secondary metabolites (Milner et al. 1996). Improvements in the plant disease management and productivity are mainly mediated through pathogen antagonism, plant growth promotion, and stimulation of defense response in host plant against the pathogen. Plant growth-promoting rhizobacteria (PGPS) suppress the growth of soilborne phytopathogens through the production of allelochemicals such as siderophores, antibiotics, and mycolytic enzymes, viz., chitinases, β-1, 3-glucanase, proteases, lipases, etc. (Whipps 2001). Rhizobacteria association with plant roots may increase plant yield through mechanisms that help in improved nutrient uptake, plant disease suppression, or production of phytohormone (Defago and Keel 1995). Plant rhizobacteria maintain a symbiotic association with the surface of plant roots (Lutenberg and Dekkers 1999). Decreased biocontrol activity may be associated with poor root colonization by rhizobacteria (Schippers et al. 1987).

2.4 Production of Secondary Metabolites by *Pseudomonas fluorescens*

Use of *P. fluorescens* has revolutionized the field of biological control in suppressing soilborne plant pathogens by producing antibiotics such as phenazine (Toohey et al. 1965), pyrrolnitrin (Burkhead and Geoghegan 1994), siderophores (Sakthivel et al. 1986), and phloroglucinol (Howell and Stipanovic 1980) that can help in controlling wilt (Fridlender et al. 1993). The biocontrol activity of *Pseudomonas* spp. is mainly mediated via the production of secondary metabolites and hydrolytic enzymes and through competitive exclusion (Elasri et al. 2001). *P. fluorescens* produce various secondary metabolites including antibiotic compounds that have been evaluated for biocontrol activity against plant pathogens mainly by genetic techniques. Antibiotics produced by *Pseudomonas* spp. inhibit metabolic activities and growth of pathogens. Antifungal secondary metabolites, viz., 2,4-diacetylphloroglucinol, pyoluteorin, phenazines, pyrrolnitrin, and HCN, contribute to the suppression of disease incidence in various host-pathogen systems. Howell and Stipanovic (1980) studied the importance of antibiotics secreted by *P. fluorescens* Pf-5 in the suppression of *Pythium ultimum* causing damping-off in cotton seedlings. Various secondary metabolites such as pyrrole-type antibiotics, phenazines, pyo-compounds (pyocyanin or pyoverdine), and indole derivatives have been characterized. Metabolites such as (amino-2-chloro-3-phenyl)-4-pyrrole-2-carboxylic acid, 7-chloroindole-3-acetic acid, and 3-chloroanthranilic acid produced by *Pseudomonas aureofaciens* at an early stage of fermentation have been reported by Salcher et al. (1978). The two-component global regulatory system GacS/GacA is known to control secondary metabolite production, viz., pyoluteorin, 2,4-DAPG, pyrrolnitrin, phenazine, HCN, exoprotease, and chitinase compound as well as siderophores (Chin-A-Woeng et al. 2000).

Enzymes produced by pseudomonads can lyse fungal cell walls but not plants, thereby preventing proliferation of plant pathogens. Hydrolytic enzymes, viz., chitinases, β-1,3-glucanases, lipases, proteases, etc., are produced by pseudomonads which are known to digest fungal cell walls, thus using them as an energy source (Leah et al. 1991) and thus making them as potential biocontrol agents (Garbeva et al. 2004). Synergistic effects have been observed on nodulation and plant growth of legume crops by inoculation of mixtures of *B. japonicum* and *P. fluorescens* in soybean (Li and Alexander 1988), *R. leguminosarum* and *P. fluorescens* strain F113 in pea (Andrade et al. 1998), and *Bradyrhizobium*/*Mesorhizobium* and *Pseudomonas* sp. in chickpea and green gram, respectively (Goel et al. 2000; Sindhu et al. 2002).

2.5 Mode of Action of Secondary Metabolites Produced by Pseudomonads

Biological control of plant pathogens by PGPR generally includes the production of antibiotics (Haas and Defago 2005), HCN (Dowling and O'Gara 1994), cell wall-degrading enzymes, viz., chitinase, protease, β-1-3-glucanase, and lipase, which can lyse the cell walls of the fungal pathogen (Chet and Inbar 1994). Characterizing potential biocontrol candidates against soilborne pathogens is more important for carrying out a successful action against plant pathogens in a dynamic and complex rhizosphere condition. A brief description of the mechanisms through which pseudomonads function to control plant pathogen and thus ultimately plant diseases is described herewith.

2.5.1 Through Antibiotic-Mediated Suppression of Plant Diseases

2.5.1.1 2, 4-Diacetylphloroglucinol (2,4-DAPG)

2,4-DAPG is a natural phenol specifically produced by gram-negative bacterium, i.e., *P. fluorescens*, and is responsible for its biocontrol and antiphytopathogenic properties. 2, 4-DAPG is the best-known phloroglucinol compound that includes monoacetylphloroglucinol and diacetylphloroglucinol formed by uncharacterized condensation of phloroglucinol and monoacetylphloroglucinol in a family of related molecules (Mavrodi et al. 2001). Troppens et al. (2013) proposed that 2,4-DAPG acts as a proton ionophore which dissipates the proton gradient across the mitochondrial membrane. The uncoupling of ATP synthesis and respiration ultimately leads to inhibition of plant pathogen which is the lethal effect of 2,4-DAPG.

2.5.1.2 Pyoluteorin (Plt)

Pyoluteorin is an aromatic chlorinated polyketide compound mainly produced by *P. fluorescens* and is effective against oomycetes like *Pythium ultimum.* Bender et al. (1999) isolated pyoluteorin from *P. fluorescens* Pf-5 and *P. aeruginosa* for the first time. Howell and Stipanovic (1980) reported that its antimicrobial properties and its application suppressed the cotton damping-off in cotton seeds caused by pathogen *Pythium ultimum.*

2.5.1.3 Pyoverdine (Pvd) or Siderophores

Siderophores are low-molecular-weight extracellular compounds having a high affinity for ferric ions (Fe^{3+}) and bind with Fe^{3+} ions to form a ferric-siderophore complex that cannot be utilized by the pathogen but the producing organism can use it via specific receptors in their outer cell membrane. The ability to bind Fe^{3+} ions provides a competitive advantage to microorganisms. The siderophores produced by *P. fluorescens* play an important role in the promotion of plant growth (Kloepper et al. 1980). *Pseudomonas fluorescens* is also known to produce siderophores which are fluorescent and yellowish-green water-soluble pigments under iron deficit conditions (Sullivan and Gara 1992). Moreover, *P. fluorescens* is known to produce several types of siderophores, i.e., salicylic acid, pyoverdine, and pyochelin (Dave and Dube 2000), and to control chickpea wilt by induced systemic resistance (ISR) via the production of salicylic acid (SA) as a signaling molecule in a medium as well as in the rhizosphere (Saikia et al. 2003). Induction of ISR via salicylic acid-dependent pathway in chickpea plants by *Pseudomonas* spp. via the production of phenolic compounds has been reported by Singh et al. (2003).

2.5.1.4 Phenazine (Phz)

Phenazines are nitrogen-containing heterocyclic compounds produced by *Pseudomonas* spp. Phenazines are produced by certain members of the pseudomonads that are redox agents and are toxic to competing organisms. As has been reported by Wienberg (1969), *P. fluorescens* produced phenazine derivative, i.e., PCA (phenazine-1-carboxylic acid), whereas *P. aureofaciens* produced two phenazine derivatives, i.e., PCA and 2- hydroxyphenazines. Almost all phenazine compounds exhibited a broad spectrum of antimicrobial activity against phytopathogens. *P. fluorescens* is among the first few microbes from which phenazine compounds were isolated and purified and reported to exhibit activity against fungal pathogens (Gurusiddaiah et al. 1986). It is largely unknown how pseudomonads themselves respond and survive in the presence of these compounds.

2.5.1.5 Pyrrolnitrin (Prn)

Pyrrolnitrin is an antifungal metabolite produced by members of the genus *Pseudomonas* spp. Arima et al. (1964) first described phenyl pyrrole derivative used as fungicide in agriculture. A four-gene cluster (prnABCD) responsible for pyrrolnitrin synthesis was first reported in *Pseudomonas aurantiaca* BL915, earlier identified as *Pseudomonas fluorescens* (Gross and Loper 2009).

2.5.1.6 Hydrogen Cyanide (HCN)

Hydrogen cyanide is mainly produced by plant growth-promoting rhizobacteria which plays an important role in biological control (Defago et al. 1990). HCN is weakly acidic and partially ionizes in water to give cyanide ions (CN^-). Cyanide ions from HCN interfere with the enzymes of the respiratory system and inhibit the action of cytochrome oxidase of the electron transport chain (Gehring et al. 1993). The energy supply to the cell is disrupted which leads to the death of the invading organism. It also inhibits the activity of enzymes and natural receptors via reversible inhibition (Corbett 1974). As have been reported by Voisard et al. (1989), fluorescent pseudomonads isolated from potato and wheat rhizosphere can produce HCN.

2.5.2 Through Cell Wall-Degrading Enzymes/Hydrolytic Enzymes

2.5.2.1 Chitinases

Chitinases fall into three classes, viz., endochitinases, 1,4-β-N-acetylglucosaminidases, and exochitinases or chitobiosidases, depending on the mechanism of chitin degradation (Viterbo et al. 2002). Chitin is a polymer made of N-acetyl-D-glucosamine (GlcNAc) units linked through β-1,4 glycosidic bonds which are mainly degraded by chitinases. Nandakumar et al. (2007) reported the production of chitinases by strains of *P. fluorescens*, viz., PF1, PB2, and FP7, on the addition of chitin source in culture medium and maximum chitinase (31.2%) is recorded by strain FP7. The addition of chitin results in a significant increase of chitinase activity (Nandakumar et al. 2007). *P. fluorescens* is known to have strong antimicrobial activity against *Rhizoctonia solani*, *Pyricularia oryzae*, *Xanthomonas oryzae*, and *Fusarium oxysporum* under in vitro and field conditions (Vidhyasekaran et al. 2001; Nandakumar et al. 2001). Expression of enzymes, viz., chitinases and β-1,3-glucanases, was reported in chickpea by Vogelsang and Barz (1993), and the presence of four isoforms of these enzymes in stems and roots of chickpea crop induced by wounding or by ethephon has been reported by Cabello et al. (1994).

Stevenson et al. (1994) and Stevenson et al. (1995) reported the induction of β-1,3-glucanase and chitinase activities in cell suspension of chickpea that are susceptible to *A. rabiei*. Stevenson et al. (1997) reported that root exudates of chickpea plants contain phytoalexins that play an important role in contributing resistance against *Fusarium* wilt under in vitro conditions.

2.5.2.2 Lipases

Lipase hydrolyzes triacylglyceride into fatty acids, di-acylglycerols, and mono-acylglycerols and also catalyzes esterification and trans-esterification reactions (Fernandes et al. 2007). Prasad (2014) isolated lipase-producing microorganisms from different soil samples that are rich in lipid content like oil mills, and the maximum lipase activity by the isolate *Pseudomonas aeruginosa* was reported at pH 7 at 35 °C for 45 hours.

2.5.2.3 Proteases

Proteases are enzymes that hydrolyze proteins into its constituent amino acids. These proteases are also known as proteolytic enzymes or systemic enzymes. Proteases can hydrolyze proteins as long as they are not part of living cells. Normal living cells are protected from lysis via the inhibitor mechanism. As have been reported by Giri et al. (1998), the differential expression of proteinase inhibitors and its accumulation are induced by wounding in *Helicoverpa armigera* against the production of proteases, which is not sensitive to inhibition by protease and can degrade them.

2.5.2.4 β-1,3-Glucanases

Glucanases hydrolyze the glycosidic bond in glucan, a polysaccharide of several glucose sub-units. β-1,3-Glucan commonly known as laminarin is a polymer of D-glucose that is arranged as helical coils in a β-1,3 configuration. Cell walls of fungi contain about 60% laminarin that is mainly hydrolyzed by glucanases or laminarinase (Onsori et al. 2005). Glucanases are mainly produced by microbes (fungi and bacteria) (Zhu et al. 2008). Exo-β-1,3-glucanases break glucose residues into monosaccharide from nonreducing ends, whereas endo-β-1,3-glucanases cleave polysaccharide chain into oligosaccharides at random sites (Vazquez-Garciduenas et al. 1998; Vijayendra and Kashiwagi 2009). β-1,3-Glucanases from bacterial and fungal sources are known to be involved in the degradation of polysaccharides into its constituent sub-units and used them as an energy source (Planas 2000).

Induction of phytoalexins and pathogenesis-related proteins, i.e., β-1,3-glucanases, may be associated with a reduction in disease incidence in chickpea (Kuc 2006). Under in vitro conditions, the purified chitinases and β-1,3-glucanases exhibited antifungal activity against β FOC (Saikia et al. 2005) indicating their direct effect on the pathogen growth. Harsha et al. (2012) reported the antifungal activity of glucanase enzyme produced by *P. fluorescens* and its use as biocontrol agent in agriculture.

2.5.2.5 Xylanases

Xylanases are enzymes that degrade linear polysaccharide, i.e., β-1,4-xylan, into xylose sub-units and hydrolyze hemicellulose which is the major component of plant cell walls. It helps in the degradation of plant matter into useful nutrients by microorganisms, viz., fungi, bacteria, and yeast. The filamentous fungi are the commercial source of xylanase (Beg et al. 2001).

2.5.3 Production of Plant Growth-Promoting Substances (PGPS)

2.5.3.1 Indole Acetic Acid (IAA)

Indole-3-acetic acid is a naturally occurring phytohormone (auxins) and is commonly produced by plant growth-promoting rhizobacteria (Barazani and Friedman 1999). IAA is involved in the root initiation, enlargement of the cell, and cell division (Salisbury 1994). Biofertilizing PGPR plays an important role in the production of IAA and its implications in plant growth promotion (Passari et al. 2015a; Vessey 2003). IAA is believed to enhance root growth, resulting in a greater area of the root surface, and thus helps the plants to acquire more nutrients from the rhizosphere.

Barea et al. (1976) isolated bacteria from the rhizosphere which can produce IAA, gibberellins, and cytokinins and found that out of the total, 17 isolates belong to the *Pseudomonas* spp. Production of IAA and GA by *Pseudomonas striata* was also reported by Sattar and Gaur (1987). It has been reported that IAA production is the inherent mechanism of PGPRs like *Pseudomonas* spp. (Mazumdar et al. 2007). As have been reported by Kumar et al. (2007), *P. fluorescens* strain Pf4–99 is capable of producing IAA in culture medium and is most effective in the improvement of chickpea crops under controlled greenhouse conditions and natural field conditions. Rhizobacteria from the roots of legume crops such as pea, lentil, and chickpea are capable of producing IAA (Hynes et al. 2008).

2.5.3.2 Gibberellins (GA)

Gibberellic acid commonly known as gibberellins is a phytohormone mainly found in plants and is capable of promoting plant growth and cell elongation. It helps in the stimulation of cells of germinating seeds to produce mRNA molecules encoding for hydrolytic enzymes. It is associated with the modification of plant morphology by the elongation of plant tissue (Salisbury 1994). The evidence for the production of GA by PGPRs has been provided by Gutierrez-Manero et al. (2001). The production of plant hormones such as IAA, GA, and cytokinins by PGPRs played a direct role in plant growth promotion and also helps in nitrogen fixation (Patten and Glick 1996).

As have been reported by Siddiqui et al. (1998), *P. fluorescens* can control wilt disease in pigeon pea caused by *H. cajani* when used alone or in combination with pesticides. *Pseudomonas* spp. have the potential to increase plant growth, nodulation in leguminous plants, and phosphorus solubilization and decrease nematode multiplication, thereby suppressing wilting in infected plants. Saikia et al. (2004) found that *P. fluorescens* isolated from rhizosphere of broad bean has antagonistic activity against fungal pathogens, viz., *Rhizoctonia solani* and *Macrophomina phaseolina*, and also reported the suppression of *Fusarium* wilt and charcoal rot in chickpea by *P. aeruginosa* strain RsB29.

2.6 Role of Biocontrol Agents in Induced Systemic Resistance (ISR) and Systemic Acquired Resistance (SAR)

Induced systemic resistance is the enhanced defensive ability developed within the host plant by nonpathogenic forms of rhizobacteria (Van Loon et al. 1998). ISR in carnation plants was induced by *P. fluorescens* strain WCS417r against *F. oxysporum* f. sp. *dianthi* (Van Peer et al. 1991). In cucumber plants, it was induced by rhizobacterial strains against the anthracnose caused by *Colletotrichum orbiculare* (Wei et al. 1991). ISR mediated through rhizobacteria resembles pathogen-mediated systemic acquired resistance (SAR) that render resistance in uninfected plant parts against plant pathogens (Van Wees et al. 1997). *Bacillus* spp. and *Pseudomonas* spp. are the most widely studied rhizobacteria that induce the ISR (Van Wees et al. 2008). ISR is induced by PGPR or nonpathogenic rhizobacteria, whereas SAR is triggered by a localized infection. ISR and SAR are mediated through a different set of signaling pathways. SAR is mediated through salicylic acid (SA) pathway, whereas two signaling pathways, i.e., jasmonic acid (JA) and ethylene (ET) pathways, are involved in ISR (Van Loon et al. 1998). The defense responses are induced by these signaling molecules when they are applied exogenously (Ryals et al. 1996). ISR-mediated resistance is significantly less than that of SAR-mediated resistance (Van Loon 2000). ISR and SAR jointly provide a better resistance response which indicates that they act in coordination in inducing the resistance response against pathogens (Van Wees et al. 2000).

The high concentration of ET and JA is a sign of defense response in infected plants (Mauch et al. 1984). In *Arabidopsis*, JA and the ET response mutants (jar1 and etr1) were tested in the induction of ISR against *P. syringae* pv. tomato by Pieterse et al. (1998) and found that these mutants were unable to induce ISR-mediated resistance in tomato upon colonization of the roots by rhizobacteria WCS417r. Methyl jasmonate (MeJA) and the ethylene precursor 1-aminocycloprop

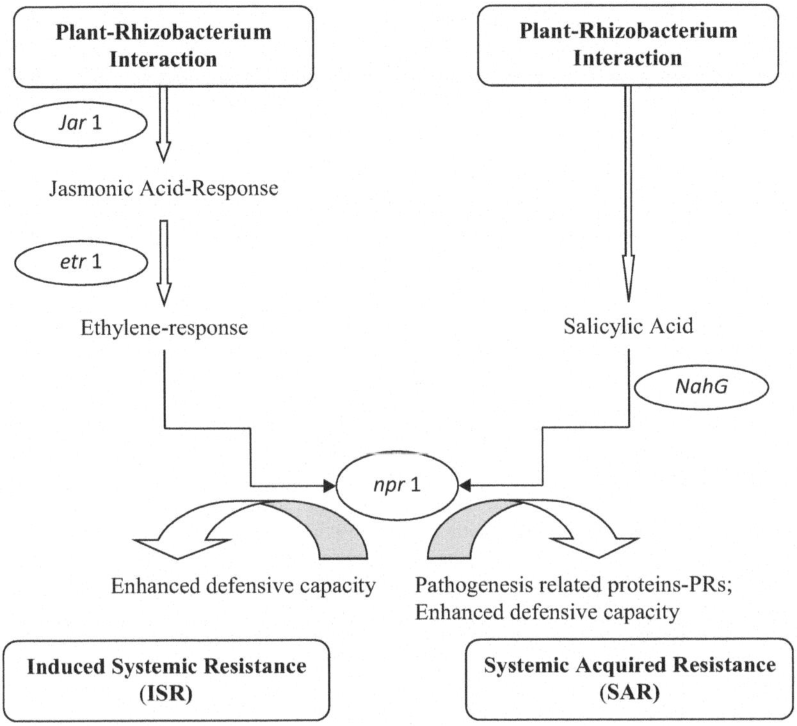

Fig. 2.1 Signal transduction pathways leading to rhizobacteria-mediated induced systemic resistance (ISR) and pathogen induced systemic acquired resistance (SAR) in *Arabidopsis thaliana*. (Source: Modified from: Van Loon et al. 1998).

ane-1-carboxylate (ACC) promote resistance against *P. syringae* pv. tomato in salicylic acid non-accumulating (NahG) plants. MeJA-mediated resistance is blocked in etr1–1, npr1–1, and jar1–1 plants, while ACC-mediated resistance is affected in npr1–1 and etr1–1 plants, but not in jar1–1 plants. Thus, WCS417r-mediated ISR follows JA- and ethylene-mediated signaling pathways, and these signaling molecules are successively coordinated to induce a defense mechanism like SAR which is regulated by NPR1 (Pieterse et al. 1998). Signal transduction pathways leading to rhizobacteria-mediated ISR and pathogen-mediated SAR in *Arabidopsis thaliana* are summarized in Fig. 2.1.

2.7 Future Perspective

The area under the legume crop cultivation and its production has not been increased in the last few years. Fungal pathogens and pests are recurrent problems for pulse crops. The chickpea pulse crop is widely grown under diverse climate conditions ranging from temperate to subtropical climates. The exploitation of the plant-microbe interaction will benefit us to identify novel secondary metabolites having antagonistic activity against disease-causing pathogens. Biological control agents are commercially available now, and these are formulated to control diseases caused by pathogens through nutrient competition and increasing resistance in plants. Biocontrol agents could be used to reduce the intensive use of agrochemicals and synthetic pesticides as they contain potential active ingredients. Thus, a strategy including the exploitation of secondary metabolites by biocontrol agents needs to be developed for integrated disease management.

References

AitBarka E, Gognies S, Nowak J, Audran JC, Belarbi A. Inhibitory effect of endophyte bacteria on *Botrytis cinerea* and its influence to promote the grapevine growth. Biol Control. 2002;24:135–42.

Alstrom S. Characteristics of bacteria from oilseed rape concerning their biocontrol activity against *Verticilliumdahliae*. J Phytopathol. 2000;149:57–64.

Andrade G, De Leij FAAM, Lynch JM. Plant mediated interactions between *Pseudomonas fluorescens*, *Rhizobium leguminosarum* and arbuscularmycorrhizae on pea. Lett Appl Microbiol. 1998;26:311–6.

Anonymous.Commodity profile of pulses – March 2015. Department of Agriculture and Co-operation, Ministry of Agriculture, Government of India; 2015.

Arima K, Imanaka H, Kausaka M, Fukuda A, Tameera C. Pyrrolnitrin a new antibiotic substance, produced by *Pseudomonas*. Agric Biol Chem. 1964;28:575–6.

Barazani O, Friedman J. Is IAA the major root growth factor secreted from plant growth mediated bacteria? J Chem Ecol. 1999;25:2397–407.

Barea JM, Navapro E, Montoya E. Production of plant growth regulators by rhizosphere phosphate solubilizing bacteria. J Appl Bacteriol. 1976;40:129–34.

Beg QK, Kapoor M, Mahajan L, Hoondal GS. Microbial xylanases and their industrial applications: a review. Appl Microbiol Biotechnol. 2001;56:326–38.

Bender CL, Rangaswamy V, Loper JE. Polyketide production by plant-associated Pseudomonads. Annu Rev Phytopathol. 1999;37:175–96.

Burkhead K, Geoghegan MJ. Antibiotics. In: Burkhead K, editor. Soil-borne Plant Pathogens. New York, NY: Macmillon; 1994.

Cabello F, Jorrin JV, Tena M. Chitinase and β-1,3-glucanase activities in chickpea (*Cicer arietinum*). Induction of different isoenzymes in response to wounding ethephon. Physiol Plant. 1994;92:654–60.

Calvo AM, Wilson RA, Bok JW, Keller NP. Relationship between secondary metabolism and fungal development. Microbiol Mol Biol Rev. 2002;66:447–59.

Chen C, Bauske EM, Musson G, Rodriguez-Kabana R, Kloepper JW. Biological control of *Fusarium* wilt on cotton by use of endophytic bacteria. Biol Control. 1995;5:83–91.

Chet I, Inbar J. Biological control of fungal pathogens. Appl Biochem Biotechnol. 1994;48:37–43.

Chin-A-woeng TF, Bloemberg GV, Mulders IH, Dekkers LC, Lugtenberg BJ. Root colonialization by Phenazine-1-Carboxamide producing bacterium *Pseudomonas chlororaphis* PCL 1391 is essential for biocontrol of tomato foot and root rot. Mol Plant-Microbe Interact. 2000;13:1340–5.

Compant S, Reiter B, Sessitsch A, Nowak J, Clement C, Barka EA. Endophytic colonization of *Vitis vinifera* L. by plant growth-promoting bacterium *Burkholderia* sp. strain PsJN. Appl Environ Microbiol. 2005;71:1685–93.

Corbett JR. The Biochemical mode of action of pesticides. London: Academic Press; 1974. p. 330.

Dave BP, Dube H. Detection and chemical characterisation of siderophores of rhizobacterial fluorescent pseudomonads. Indian Phytopathol. 2000;53:97–8.

Defago G, Berling CH, Burger U, Hass D, Hahr G, Keel C, Voisard C, Wirthner PH, Wutrich B. Suppression of black root rot of tobacco by a *Pseudomonas* strain: potential applications and mechanisms. In: Hornby D, editor. Biological control of soil-borne plant pathogens. Oxfordshire: CAB International; 1990. p. 93–108.

Defago G, Keel C. *Pseudomonads* as biocontrol agents of diseases caused by soil borne pathogens. In: HMT H, Lynch JM, editors. Benefits and risks of introducing biocontrol agents. Cambridge, UK: University Press; 1995.

Deshwal VK, Pandey P, Kang SC, Maheshwari DK. Rhizobia as biological control agent against soil-borne plant pathogenic fungi. Ind J Exp Biol. 2003;41:1160–4.

Dowling DN, O'Gara F. Metabolites of *Pseudomonas* involved in the biocontrol of plant disease. Trends Biotechnol. 1994;12:133–41.

Elasri M, Delorme S, Lemanceau P, Stewart G, Laue B, Glickmann E, Oger PM, Dessaux Y. Acyl – homoserine lactone production is more common among plant-associated *Pseudomonas spp.* than among soil borne *Pseudomonas* spp. Appl Environ Microbiol. 2001;67:1198–209.

Fernandes MLM, Saad EB, Meira JA, Ramos LP, Mitchel DA, Kriegeir N. Esterification and transesterification reactions catalysed by addition of fermented solids to organic reaction media. J Mol Catal B Enzym. 2007;44:8–13.

Fridlender M, Inbar J, Chet I. Biological control of soilborne pathogens by a β-13 glucanase producing *Pseudomonas cepacia*. Soil Biol Biochem. 1993;25:1211–21.

Garbeva P, van Vean JA, van Elas JD. Assessment of the diversity and antagonism towards *Rhizoctonia solani*AG3 of *Pseudomonas* spp. in soil from different agricultural regimes. FEMS Microbiol Ecol. 2004;47:51–64.

Gehring PJ, Nolan RJ, Watanabe PG. Solvents, fumigants and related compounds. In: Hayes WJ, Laws ER, editors. Handbook of pesticide toxicology, vol. 2. San Diego, CA: Academic Press Inc.; 1993. p. 646–9.

Giri AP, Harsulkar AM, Deshpande VV, Sainani MN, Gupta VS, Ranjekar PK. Chickpea defensive proteinase inhibitors can be inactivated by podborer gut proteinases. Plant Physiol. 1998;116:393–401.

Goel AK, Sindhu SS, Dadarwal KR. Pigment diverse mutants of *Pseudomonas* sp.: inhibition of fungal growth and stimulation of growth of *Cicer arietinum*. Biol Plant. 2000;43:563–9.

Gohain A, Sarma RK, Debnath R, Saikia J, Singh BP, et al. Phylogenetic affiliation and antimicrobial effects of endophytic actinobacteria associated with medicinal plants: prevalence of polyketide synthase type II in antimicrobial strains. Folia Microbiol. 2019;64(4):481–96.

Gopalakrishnan S, Pande S, Sharma M, Humayun P, Kiran BK, Sandeep D, Vidya MS, Deepthi K, Rupela O. Evaluation of actinomycete isolates obtained from herbal vermicompost for the biological control of *Fusarium* wilt of chickpea. Crop Prot. 2011;30:1070–8.

Gross H, Loper, JE. Genomics of secondary metabolite production by *Pseudomonas* spp. Nat Prod Rep. 2009;26:1408–46.

Gurusiddaiah S, Weller DM, Sarkar A, Cook RJ. Characterisation of an antibiotic produced by an strain of *Pseudomonas fluorescens* inhibitory to *Gaeumannomyces graminis* var. tritici and *Pythium* spp. Antimicrob Agents Chemother. 1986;29:488–95.

Gutierrez-Manero FJ, Ramos-Solano B, Probanza A, Mehouachi J, Tadeo FR, Talon M. The plant growth promoting rhizobacteria *Bacillus pumilus* and *B. licheniformis* produce high amounts of physiologically active gibberellins. Plant Physiol. 2001;111:206–11.

Haas D, Defago G. Biological control of soil borne pathogens by fluorescent pseudomonads. Nat Rev Microbiol. 2005;3:307–19.

Hallmann J, Quadt AV, Mahaffee WF, Kloepper J. Endophytic bacteria in agricultural crops. Can J Microbiol. 1997;43:895–914.

Harsha P, Sasidharan N, Madhavan K, Venkatachalam D, Rajendran S, Thayumanavan T, Babu S. Comparative evaluation of rice and sunhemp root inhabiting *Pseudomonas fluorescens* for optimized glucanase production. J Agric Biotech Sustain Dev. 2012;4:50–6.

Howell CR, Stipanovic RD. Suppression of *Pythium ultimum* induced damping-off of cotton seedlings by *Pseudomonas fluorescens* and its antibiotic, pyoluteorin. Phytopathology. 1980;70:712–5.

Hynes RK, Leung GCY, Hirkala DLM, Nelson LM. Isolation, selection and characterization of beneficial rhizobacteria from pea, lentil and chickpea grown in Western Canada. Can J Microbiol. 2008;54:248–58.

Kaur NP, Mukhopadhayay AN. Integrated control of chickpea wilt complex by Trichoderma spp. and chemical methods in India. Trop Pest Manag. 1992;38:372–5.

Kloepper JW, Leong J, Teintze M, Schroth MN. Enhanced plant growth by siderophores produced by plant growth-promoting rhizobacteria. Nature. 1980;286:885–6.

Kravchenko LV, Makarova NM, Azarova TS, Provorov NA, Tikhonovich IA. Isolation and phenotypic characteristics of growth-stimulating rhizobacteria (PGPR), with high root- colonizing and phytopathogenic fungi inhibiting abilities. Microbiol. 2002;71:521–5.

Kuc J. What's old and what; new in concepts of induced systemic resistance in plants, and its applications. In: Tuzun S, Bent E, editors. Multigenic and induced resistance in plants. New York, NY: Springer; 2006. p. 9–20.

Kumar H, Bajpai VK, Dubey RC, Maheshwari DK, Kang SC. Wilt disease management and enhancement of growth and yield of *Cajanus cajan* (L) var. Manak by bacterial combinations amended with chemical fertilizer. Crop Prot. 2010;29:591–8.

Kumar V, Kumar A, Kharwar RN. Antagonistic potential of fluorescent pseudomonads and control of charcoal rot of Chickpea caused by *Macrophomina phaseolina*. J Environ Biol. 2007;28:15–20.

Leah R, Tommerup S, Svendsen I, Murphy J. Biochemical molecular characterization of three barley seed proteins with antifungal properties. J Biol Chem. 1991;266:1564–73.

Li DM, Alexander M. Co-inoculation with antibiotic-producing bacteria to increase colonization and nodulation by rhizobia. Plant Soil. 1988;108:211–9.

Liu L, Kloepper JW, Tuzun S. Induction of systemic resistance in cucumber against *Fusarium* wilt by plant growth- promoting rhizobacteria. Phytopathology. 1995;85:695–8.

Lutenberg BJ, Dekkers LC. What makes *Pseudomonas* bacteria rhizosphere competent? Environ Microbiol. 1999;1:9–13.

Mahajan K, Sharma JK, Dhage A. Evaluation of *Trichoderma* sp. against *Fusarium* wilt of chickpea caused by *Fusarium oxysporum* f. sp. *ciceris* under in vitro condition. Int J Curr Microbiol Appl Sci. 2018;7:595–99.

Mauch F, Hadwiger LA, Boller T. Ethylene: symptom, not signal for the induction of chitinase and-1,3-glucanase in pea pods by pathogens and elicitors. Plant Physiol. 1984;76:607–11.

Mavrodi OV, McSpadden GBB, Mavrodi DV, Bonsall RF, Weller DM, Thomashow LS. Genetic diversity of *phlD* from 2, 4-diacetylphloroglucinol-producing fluorescent *Pseudomonas* spp. Phytopathology. 2001;91:35–43.

Mazumdar T, Goswami C, Talukdar NC. Characterization and screening of beneficial bacteria obtained on King's B agar from tea rhizosphere. Indian J Biotechnol. 2007;6:490–4.

Meki S, Ahmed S, Sakhuja PK. Control of chickpea wilt (*Fusarium oxysporum* f. sp. *ciceris*) using *Trichoderma* spp. in Ethiopia. Arch Phytopathol Plant Protect. 2009;44:432–40.

Milner JL, Silo-Suh L, Lee JC, He H, Clardy J, Handelsman J. Production of kanosamine by *Bacillus cereus* UW85. Appl Environ Microbiol. 1996;62:3061–5.

Mishra VK, Passari AK, Singh BP. In vitro antimycotic and biosynthetic potential of fungal endophytes associated with Schima wallichii. In: Kumar P, et al., editors. Current trends in plant disease diagnostics and management practices, fungal biology. Cham: Springer International Publishing Switzerland; 2016. p. 367–81.

Nandakumar R, Babu S, Viswanathan R, Raguchander T, Samiyappan R. Induction of systemic resistance in rice against sheath blight disease by *Pseudomonas fluorescens*. Soil Biol Biochem. 2001;33:603–12.

Nandakumar R, Babu S, Raguchander T, Samiyappan R. Chitinolytic activity of native *Pseudomonas fluorescens* strains. J Agr Sci Technol. 2007;9:61–8.

Nautiyal CS. Biocontrol of plant diseases for agricultural sustainability. In: Upadhyay RK, Mukerji KG, Chamola BP, editors. Biocontrol potential and its exploitation in sustainable agriculture, Crop Diseases, Weeds, and Nematodes, vol. 1. New York: Kluwer Academy Plenum; 2000. p. 9–23.

Oh DC, Jensen PR, Kauffman CA, Fenical W, Libertellenones A-D. Induction of cytotoxic diterpenoid biosynthesis by marine microbial competition. Bioorg Med Chem. 2005;13:5267–73.

Onsori H, Zamani R, Motallebi M, Zorghami N. Identification of over producer strain endo-β-1, 4-glucanase in *Aspergillus* species: characterization of crude carboxymethylcellulase. Afr J Biotechnol. 2005;4:26–30.

Passari AK, Chandra P, Zothanpuia, Mishra VK, Leo VV, et al. Detection of biosynthetic gene and phytohormone production by endophytic actinobacteria associated with *Solanum lycopersicum* and their plant-growth-promoting effect. Res Microbil. 2016;167:692–705.

Passari AK, Lalsiamthari PC, Zothanpuia, Leo VV, Mishra VK, et al. Biocontrol of *Fusarium* wilt of *Capsicum annuum* by rhizospheric bacteria isolated from turmeric endowed with plant growth promotion and disease suppression potential. Eur J Plant Pathol. 2017;150:831–46.

Passari AK, Mishra VK, Saikia R, Gupta VK, Singh BP. Isolation, abundance and phylogenetic affiliation of endophytic actinomycetes associated with medicinal plants and screening for their *in vitro* antimicrobial biosynthetic potential. Front Microbiol. 2015a;6:273, pp. 1–18.

Passari AK, Mishra VK, Gupta VK, Yadav MK, Saikia R, et al. *In Vitro* and *In Vivo* plant growth promoting activities and DNA fingerprinting of antagonistic endophytic actinomycetes associates with medicinal plants. PLoS ONE. 2015b;10(9):e0139468, pp. 1–18.

Passari AK, Upadhyaya K, Singh G, Abdel-Azeem AM, Thankappan S, et al. Enhancement of disease resistance, growth potential, and photosynthesis in tomato (*Solanum lycopersicum*) by inoculation with an endophytic actinobacterium, *Streptomyces thermocarboxydus* strain BPSAC147. PLoS One. 2019;14(7):e0219014.

Patten C, Glick BR. Bacterial biosynthesis of Indole-3-acetic acid. Can J Microbiol. 1996;42:207–20.

Pavlo A, Leonid O, Iryna Z, Natalia K, Maria PA. Endophytic bacteria enhancing growth and disease resistance of potato (*Solanum tuberosum* L.). Biol Control. 2011;56:43–9.

Pieterse CMJ, Van Wees SC, Van Pelt JA, Knoester M, Laan R, Gerrits H, Weisbeek PJ Van Loon LC. A novel signaling pathway controlling induced systemic resistance in *Arabidopsis*. Plant Cell. 1998;10:1571–80.

Planas A. Bacterial 1,3-1,4-beta-glucanase: structure, function and protein engineering. Biochem Biophys Acta. 2000;1543:361–82.

Pleban S, Ingel F, Chet I. Control of *Rhizoctonia solani* and *Sclerotium rolfsii* in the greenhouse using endophytic *Bacillus* spp. Eur J Plant Pathol. 1995;101:665–72.

Prasad MP. Production of Lipase enzyme from *Pseudomonas aeruginosa* isolated from lipid rich soil. Int J Pure App Biosci. 2014;2:77–81.

Ryals JA, Neuenschwander UH, Willits MG, Molina A, Steiner H-Y, Hunt MD. Systemic acquired resistance. Plant Cell. 1996;8:1808–19.

Safiyazov JS, Mannanov RN, Sattarova RK. The use of bacterial antagonists for the control of cotton diseases. Field Crops Res. 1995;43:51–4.

Saikia R, Singh BP, Arora DK. Detection of pathogenesis-related proteins chitinase and β-1,3-glucanase in induced chickpea. Curr Sci. 2005;89:659–63.

Saikia R, Singh K, Arora DK. Suppression of *Fusarium*-wilt and charcoal rot of chickpea by *Pseudomonas aeruginosa* RsB29. Indian J Microbiol. 2004;44:181–4.

Saikia R, Singh T, Kumar R, Srivastava J, Srivastava AK, Singh K, Arora DK. Role of salicylic acid in systemic resistance induced by *Pseudomonas fluorescens* against *Fusarium oxysporum* f. sp. ciceri in chickpea. Microbiol Res. 2003;158:203–13.

Sakthivel N, Sivamani E, Unnmalai N, Gananamanickam SS. Plant growth promoting rhizobacterial in enhancing plant growth and suppressing plant pathogens. Curr Sci. 1986;55:22–5.

Salcher O, Lingens F, Fischer P. Biosynthesis von pyrrolnitrin. Tetrahedron Lett. 1978;34:3097–100.

Salisbury FB. The role of plant hormones in plant environment interactions. New York: Marcel Dekker; 1994.

Sattar MA, Gaur AC. Production of auxins and gibberellins by phosphate dissolving microorganisms. Zentralbl Microbiol. 1987;142:393–5.
Schippers B, Bakker AW, Bakker PAHM. Interactions of deleterious and beneficial rhizosphere microorganisms and the effect of cropping practices. Annu Rev Phytopathol. 1987;25:339–58.
Siddiqui ZA, Irshad, Mahmood I, Hayat S. Biocontrol of *Heterodera cajani* and *Fusarium udum* on pigeon pea using *Glomus mosseae*, *Paecilomyces lilacinus* and *Pseudomonas fluorescens*. Thai J Agric Sci. 1998;31:310–21.
Sindhu SS, Suneja S, Goel AK, Parmar N, Dadarwal KR. Plant growth promoting effects of *Pseudomonas* sp. on coinoculation with *Mesorhizobium* sp. *Cicer* strain under sterile and "wilt sick" soil conditions. Appl Soil Ecol. 2002;19:57–64.
Singh BP, Rateb ME, Rodriguez-Couto S, Polizeli MLTM, Li W-J. Editorial: microbial secondary metabolites: recent developments and technological challenges. Front Microbiol. 2019;10:914.
Singh UP, Sarma BK, Singh DP. Effect of plant growth promoting rhizobacteria and culture filtrate of *Sclerotium rolfsii* on phenolic and salicylic acid contents in chickpea (*Cicer arietinum*). Curr Microbiol. 2003;46:131–40.
Stevenson PC, Padgham DE, Haware MP. Root exudates associated with the resistance of four chickpea cultivars (*Cicer arietinum*) to two races of *Fusarium oxysporum* f. sp. *ciceris*. Plant Pathol. 1995;44:686–94.
Stevenson PC, Padgham DE, Haware MP. The chemical basis of resistance in chickpea to *Fusarium* wilt. Acta Hortic. 1994;381:631–7.
Stevenson PC, Turner HQ, Haware MP. Phytoalexin accumulation in the roots of chickpea (*Cicer arietinum* L.) seedlings associated with resistance to *Fusarium* wilt (*Fusarium oxysporum* f. sp. *ciceris*). Physiol Mol Plant Pathol. 1997;50:167–78.
Sturz AV, Christie BR, Matheson BG, Arsenault WJ, Buchanan NA. Endophytic bacterial communities in the periderm of potato tubers and their potential to improve resistance to soil-borne plant pathogens. Plant Pathol. 1999;48:360–9.
Sullivan DJ, Gara F. Traits of fluorescent pseudomonads involved in suppression of plant root pathogens. Microbiol Rev. 1992;56:662–76.
Toohey JI, Netson CD, Krotkov G. Isolation and identification of two phenazines from a strain of *Pseudomonas aureofaciens*. Can J Bot. 1965;43:1055–62.
Troppens DM, Dmitriev RI, Papkovsky DB, O'Gara F, Morrissey JP. Genome-wide investigation of cellular targets and mode of action of the antifungal bacterial metabolite 2,4-diacetylphloroglucinol in *Saccharomyces cerevisiae*. FEMS Yeast Res. 2013;13:322–34.
Van Emden HF, Ball SL, Rao MR. Pest disease and weed problems in pea lentil and faba bean and chickpea. In: World crops: cool season food legumes, ISBN 90-247-3641-2. Dordrecht: Kluwer Academic Publishers; 1988.
Van Loon LC, Bakker PA, Pieterse CM. Systemic resistance induced by rhizosphere bacteria. Annu Rev Phytopathol. 1998;36:453–83.
Van Loon LC. Systemic induced resistance. In: Slusarenko AJ, Fraser RSS, Van Loon LC, editors. Mechanisms of resistance to plant diseases. Dordrecht: Kluwer Academic Publishers; 2000. p. 521–74.
Van Peer R, Niemann GJ, Schippers B. Induced resistance and phytoalexin accumulation in biological control of *Fusarium* wilt of carnation by *Pseudomonas* sp. Strain WCS417r. Phytopathology. 1991;81:728–34.
Van Wees SC, Van der Ent S, Pieterse CM. Plant immune responses triggered by beneficial microbes. Curr Opin Plant Biol. 2008;11:443–8.
Van Wees SCM, De Swart EAM, Van Pelt JA, Van Loon LC, Pieterse CMJ. Enhancement of induced disease resistance by simultaneous activation of salicylate- and jasmonate-dependent defense pathways in *Arabidopsis thaliana*. Proc Natl Acad Sci USA. 2000;97:8711–6.
Van Wees SCM, Pieterse CMJ, Trijssenaar A, Van'tWestend YAM, Hartog F, Van Loon LC. Differential induction of systemic resistance in *Arabidopsis* by biocontrol bacteria. Mol Plant-Microbe Interact. 1997;10:716–24.
Vazquez-Garciduenes S, Morales CAL, Estrella AH. Analysis of β-1,3glucanolytic system of biocontrol agent *Trichoderma harzianum*. Appl Environ Microbiol. 1998;64:1442–6.

Vessey KJ. Plant growth promoting rhizobacteria as biofertilizers. Plant Soil. 2003;255:571–86.
Vidhyasekaran P, Kamala N, Ramanathan A, Rajappan A, Paranidhran V, Velazhahan R. Induction of systemic resistance by *Pseudomonas fluorescens* Pf_1 against *Xanthomonas oryzae* pv. *Oryzae* in rice leaves. Phytoparasitica. 2001;29:155–66.
Vijayendra SV, Kashiwagi Y. Characterization of a new acid stable exo-beta-1,3-glucanase of *Rhizoctonia solani* and its action on microbial polysaccharides. Int J Biol Macromol. 2009;44:92–7.
Viterbo A, Ramot O, Chemin L, Chet I. Significance of lytic enzymes from *Trichoderma* spp. in the biocontrol of fungal plant pathogens. Antonie Van Leeuwenhoek. 2002;81:549–56.
Vogelsang R, Barz W. Purification characterisation and differential hormonal regulation of a β-1,3-glucanase and two chitinases from chickpea (*Cicer arietinum* L.). Planta. 1993;189:60–9.
Voisard C, Keel O, Haas P, Defago G. Cyanide production by *Pseudomonas fluorescens* helps to suppress black root rot of tobacco under gnotobiotic condition. Eur Microbiol J. 1989;8:351–8.
Wei G, Kloepper JW, Tuzun S. Induction of systemic resistance of cucumber to *Colletotrichum orbiculare* by select strains of plant growth-promoting rhizobacteria. Phytopathology. 1991;81:1508–12.
Whipps JM. Microbial interactions and biocontrol in the rhizosphere. J Exp Bot. 2001;52:487–511.
Wienberg ED. Biosynthesis of secondary metabolites: roles of trace elements. Adv Microb Physiol. 1969;4:1–44.
Yadav J, Verma JP, Tiwari KN. Plant growth promoting activities of fungi and their effect on chickpea plant growth. Asian J Biol Sci. 2011;4:291–9.
Yang J, Kloepper JW, Ryu CM. Rhizosphere bacteria help plants tolerate abiotic stress. Trends Plant Sci. 2009;14:1–4.
Zhu BW, Zhao JG, Yang JF, Mikiro T, Zhang ZS, Zhou DY. Purification and partial characterization of a novel β-1,3-glucanase from gut of a sea cucumber *Stichopus japonicus*. Process Biochem. 2008;43:1102–6.

Chapter 3
Integrated Fungal Foliar Diseases of Arid Legumes: Challenges and Strategies of Their Management in Rain-Fed Areas

Udaya Kumar Vandana, P. Bijoya Singha, Sharmista Chakraborthy, and P. B. Mazumder

3.1 Introduction

Grain legumes play a major role in improving food and nutritional security of farmers and populations, covering up to 45% of arid and semiarid regions of the world (Sprent and Gehlot 2010). Some of the globally important grain legumes which are grown worldwide and economically important are chickpea (*Cicer arietinum* L.), lentil (*Lens culinaris* Medik), cowpea (*Vigna unguiculata* (L.) Walp), and faba bean (*Vicia faba* L.) (Cernay et al. 2016; Raseduzzaman and Jensen 2017). These legumes are severely damaged by numerous plant pathogens from bacteria to fungi and viruses to nematodes causing economic losses globally (Jones et al. 2013). Among these pathogens, fungi are the largest group that affects all parts of the plants, majorly foliar parts. Fungal foliar diseases such as *Ascochyta* blight (*Ascochyta rabiei*) and *Botrytis* gray mold (*Botrytis cinerea*) affect chickpea (*Cicer arietinum*). In lentils, *Ascochyta* blight is caused by *Ascochyta lentis* and rust is caused by *Uromyces viciae-fabae* Pers. Anthracnose (*Colletotrichum lindemuthianum* Sacc. & Magn.) and *Cercospora* leaf spot (*Cercospora canescens* Fellis & Martin and *Cercospora cruenta* Sacc.) affect cowpea, respectively. Chocolate leaf spot (*Botrytis fabae* and *B. cinerea*) and rust (*Uromyces viciae-fabae*) affect faba bean (Girish et al. 2019). Challenges in sustainable management are lack of understanding of integrated pest management while adopting biopesticides in underdeveloped countries conquer the disease and are not effective as chemical fungicides and hence the farmers are not willing to use the products (Parsa et al. 2014; Peshin et al. 2009; Vandana et al. 2017). The integrated disease management (IDM) of legumes in a particular area depends upon the genetic resistance and other components of disease management (Coakley et al. 2002; Isman 2000). IDM program lies in identifying, evaluating, merging, and locating distinct components (D'Mello et al. 1998;

U. K. Vandana · P. B. Singha · S. Chakraborthy · P. B. Mazumder (✉)
Department of Biotechnology, Assam University, Silchar, Assam, India

B. P. Singh et al. (eds.), *Management of Fungal Pathogens in Pulses*, Fungal Biology, https://doi.org/10.1007/978-3-030-35947-8_3

Nel et al. 2007). This chapter emphasizes on the globally important arid and semi-arid legumes, affected by important fungal foliar diseases, and strategies of IDM for the control of fungal diseases. Approaches to sustainable management including cultural and physical practices, exploitation of host resistance, and protection with a synthetic fungicide are also discussed in the chapter.

3.2 Chickpea

Chickpea is a staple grain legume, the most prevalent food legume in the world. It serves as a major source of human diets rich in nutrients (protein) and high-quality crop residues for animal feed as well. Some of the crucial facets of chickpea are to maintain the fertility of soil via biological nitrogen fixation, furthermore in contributing to the sustainability of cropping structures by approaching practice like cereal-legume rotations. The significance of chickpea among temperate pulses is its tolerance to heat and drought in low fertility soils. Some of the important diseases affecting chickpea crop are:

3.2.1 Ascochyta *Blight (*Ascochyta rabiei*)*

Ascochyta rabiei comes under the most devastating fungal diseases of chickpea in numerous countries (Pande et al. 2005), favoring disease development and spread particularly by environmental conditions (cool and wet weather).

3.2.1.1 Diagnosis and Epidemiology

The fungal pathogen outbreaks parts above the ground of the plant. Fungi thrive on infected seeds, crop residues, and volunteer seedlings starting from one growing season to the next. When conditions are favorable and the prime source of inoculum is a seed, some dark brown lesions develop in the stem. When it comes to the air-borne spores, initial indications emerge as tiny necrotic specks on aerial parts of the primordial leaves. These specks under cool and wet conditions rapidly become enlarged and cohere, with the blighted portions having pycnidia formed all over the plant. In a susceptible culture, the necrosis progressively grows down, thereby killing the infected plant. Lesions are inversely ovate to extend and bear pycnidia on the stems and petioles. Generally, there is a breakage in stems and petioles due to engirdle. The round lesions develop on pods with some pycnidia, generally arranged in concentric rings, where the pod wall is penetrated by a fungus, infecting the seed on which lesions develop. Crop infection may emerge from seed-borne inoculum and from conidia of rain-splashed or windborne ascospores from infested parts. It was displayed that the teleomorph (the sexual reproductive stage of any fungus of phyla

Ascomycota and *Basidiomycota*) has a crucial portion in the epidemiology of the infection and played important role in controlling the disease in Spain and the United States (Kaiser et al. 2011).

3.2.1.2 Control

Disease control can include approaches such as burying the harvest debris, abolition of seed-borne inoculum, and establishing disease-resistant varieties. ICARDA and ICRISAT released numerous blight-resistant cultivars (Nene et al. 2011) which involve methods such as seeding blight-free seed, application of foliar fungicides and seed treatments and rotation of crop for 3 years, controlling diseased debris, and finally implanting blight-resistant varieties.

3.2.2 *Botrytis Gray (*Botrytis cinerea*)*

Botrytis grey mould is the common plant diseases in India, Nepal, Pakistan, and Bangladesh which is caused by *Botrytis cinerea*, which is reported to reduce yields in Australia and Argentina as well (Pande et al. 2006). Favorable conditions for the pathogen can substantially lead to major yield loss (Rashid et al. 2014).

3.2.2.1 Diagnosis and Epidemiology

A minimum of five diverse pathogen types of *B. cinerea* were identified (Kaiser et al. 2011). Furthermore, studies in pathogenic variability are mandatory. The inceptions of infection take place in the lower portions of the infected plant initially and later, under favorable condition, extend to the upper leaves. Often, there is a development of soft rot, and fungus sporulation can be noticed at the plant basal part in the seedlings which were seed infested with *B. cinerea.* Plant parts cultivated symptoms like dark-colored lesions mainly shielded with moldy fungal development. Changes such as complete engirdling of stems by lesions and breaking off of tender branches at the site where gray mold causes decomposition can be observed. Damaged leaves and flowers eventually turn into a decaying mass, and pods almost disappear or left with less quantity, withered spores (having lost all moisture). Immature seeds develop grayish-white mycelium. *B. cinerea* has a broad range of host, there is almost always a presence of the inoculum in the environment, and it can survive with other crops and weeds. Kaiser et al. (2011) conducted some experiments in a glasshouse, where they found that the fungus is being potential on 8 different crop species and 21 weed types. Feasibility of seed-borne source greatly reduces when kept in room storage. However, there is a prompt diminution in the sustainability of the fungus throughout stowage. The disease is mainly favored by moist and moderate temperatures. The respective significance of seed-borne inoculum and additional causes needs to be explored in different parts.

3.2.2.2 Control

Voluminous lines of chickpea with moderate resistance to gray mold were found although lines of resistance with increased levels have not been found. They found 22 lines with valuable resistance out of 8500 accessions evaluated. Despite a huge degree of flower infection, numerous chickpea lines produce good yields (Kaiser et al. 2011). The severity of gray mold can be reduced by the late sowing of chickpeas, but it leads to reduced yields in normal years (Kaiser et al. 2011). Gray mold can be efficiently reduced by seed treatment trialed by three sprays of carbendazim (Kaiser et al. 2011). The effectiveness of foliar sprays of vinclozolin was reported as well (Kaiser et al. 2000). Seed treatment with the spraying of triadimefon, carbendazim + thiram, mancozeb, or triadimefon was useful in checking seed-borne infection (˃94%) (Kaiser et al. 2011) followed by observation 50 days post-sowing or at the advent of indications which resulted in comprehensive control of both primary and secondary infections. However, at present, disease resistance at a high level is not found in chickpea cultivars. Therefore, moderately resistant cultivars are necessary to be developed in combination with an integrated disease management program with critical chemical use, and improved cultural procedures appear to minimize crop loss devastated by gray mold.

3.3 Lentil

Lentil is regarded as one of the important legumes considering its nutritive value. It is an outstanding source of molybdenum and folate and also serves as a rich source of copper, phosphorus, manganese, and dietary fiber (Hall et al. 2017). It serves as a staple food in countries like India, Canada, Turkey, the United States, and Nepal. According to the USDA National Nutrient Database, 353 calories can be produced from 100 g of raw lentil (Agriculture 2014). Lentils are rich in water (8%), carbohydrates (63%), dietary fiber (11%), protein (25%), and fat (1%). They are also rich in phosphorus (40% DV), iron (50% DV), zinc (35% DV), folate (120% DV), thiamin (76% DV), pantothenic acid (43% DV), and vitamin (42% DV (Faris et al. 2013).

*3.3.1 Rust (*Uromyces viciae-fabae *Pers.)*

One of the serious diseases of lentils is caused by rust (*Uromyces fabae*), which is particularly damaging the crops in countries like India, Chile, Pakistan, Ethiopia, Morocco, and Ecuador (Kaiser et al. 2011).

3.3.1.1 Diagnosis and Epidemiology

Environmental conditions (temperatures varying between 20 and 22 °C and wet weather) favour the initial infection and disease development, resulting in crop loss. All the green plants, including plant parts and pods, are infected. Early symptoms of yellowish-white pycnia (spermatogonia) and aecia (individually or in small groups) appear on the undersurface of pods and leaflets and eventually turn brown. Dark brown to black teliospores are observed to be developed on leaves and on stems and petioles. Crop genera including *Lathyrus*, *Lens*, *Pisum*, and *Vicia* are infected by the pathogen majorly. Before the establishment of a favorable and effective pathogenic relationship, there is a necessary association between the pathogenic cell surfaces and its host. Following the contact between the two faces, pre-penetration is a basic necessity for the events that lead to disease development (Negussie and Pretorius 2012). Many pathogenic fungi such as *U. viciae-fabae* produce substances that are generally present in the extracellular matrix which facilitate adhesion of gremlins and ungermlins spores.

Moreover, to extracellular matrix materials, adhesion pads of germinating urediniospores recognized to aid in the addition to the spores on the surface of the host by intensifying the part of interaction for substratum (Negussie and Pretorius 2012). The fungus thrives on infested lentil debris from season to season via teliospores. The diseased debris, when mixed with seed, became infected (Negussie and Pretorius 2012). During the growing season, aeciospores have a vital role in spreading the infection.

3.3.1.2 Control

Numerous approaches are attempted to control the disease which includes field sanitation, crop rotation, seed treatment, and use of foliar fungicides (Nene et al. 2011), and most resistance variety (Kaiser et al. 2011). ICARDA identified novel sources of rust resistance to one or more diseases by screening lentil germplasm in various parts of the world, namely, in Ethiopia, Morocco, and Pakistan, where rust epidemics are frequent. There have been several lines that have moderate resistance. Seed treatment with diclobutrazol compels in annihilating seed-borne inoculum effectively (Nene et al. 2011), and it was also reported with the efficiency of foliar sprays with mancozeb. However, some new inputs in this area of research are required to control this disease.

3.3.2 Ascochyta *Blight (*Ascochyta lentis*)*

Ascochyta blight caused by *Ascochyta lentis*, is one of the most devastating fungal diseases that restrains lentil production. It was first reported from the USSR (Nene et al. 2011).

3.3.2.1 Diagnosis and Epidemiology

A favorable environmental condition such as cool and wet weather leads to disease development and spread of *A. lentis*. It is a seed-borne disease that affects all the aboveground parts of the host plant, creating tiny, spherical gray- to dark-colored lesions along with the dark margins in the vicinity of lacerations on the leaflets. Tiny dark brown to black pycnidia appear in the abrasions on leaflets and pods. Pedersen et al. (1994) reported that although under rain-splashed condition, it leads to conidia dispersion, conditions such as wetness periods of 1–2 days will lead to infection under favorable temperature (10–15 °C). The dispersion of pathogens may also take place via wind-blown infected leaflets and seeds. The fungus thrives in crop debris. Disease epidemiology is needed to be investigated by researchers.

3.3.2.2 Control

The strategies for controlling *Ascochyta* blight in economical and sustainable ways can be via resistance breeding and cultural practices. Practices including crop rotation, early seeding for evasion of damp weather at harvest, and employing disease-free seed can be applied to minimize crop losses (Nene et al. 2011). Numerous fungicides are evaluated to control seed-borne infection with thiabendazole, benomyl, carbathin, and carbendazim having effective manifold degrees.

3.4 Cowpea

Cowpea (*Vigna unguiculata*) is a widely adapted legume. Cowpea has important nutritional content; thus, it is widely consumed by millions of people. The crop is cultivated in warm regions of the world on around seven million hectares (Adebanjo and Bankole 2004). Cowpea is produced in Asia, in North America (southeastern and southwestern regions), and largely in semiarid northeastern Brazil.

*3.4.1 Anthracnose (***Colletotrichum lindemuthianum** *Sacc. & Magn)*

Cowpea is prone to outbreak by several pathogens such as anthracnose from seeding to harvest affected by *Colletotrichum lindemuthianum* (Saccardo and Magnus) Briosi and Cavara, which is first recorded in Nigeria in 1969 (Adebanjo and Bankole 2004). Anthracnose causes a 50% yield loss in cowpea under wet and damp conditions in the regions ranging from Nigerian rainforest belt to other parts of Nicaragua; Eastern, Western, and Southern Africa; and Brazil (Williams 1975).

3.4.1.1 Diagnosis and Epidemiology

The disease is prompted to spread under cool, wet weather and particularly damage monocropped cowpeas and affect all aboveground plant parts. Individual lesions vary in shape, generally, from biconvex to circular, and color, turning tan to dark. Lines with high susceptibility can develop lesions that spread largely in number, rapidly leading to coalescing stems and twigs and petioles engirdle. Later, they appear almost completely brown. Resistant lines appear to have relatively small narrow lesions than hypersensitive lines which range from tiny necrotic flecks to shiny reddish-brown lenticular lesions of 5 mm long without sporulation. About 40% of the pathogen is seed-borne in cowpea (Adebanjo and Bankole 2004). Reduction of 35–50% in grain yield of a highly susceptible line has been measured in a monocrop culture when introduced with the disease at an initial stage during crop growth (Adebanjo and Bankole 2004). Nonetheless, the disease breakthrough is taking a relatively prolonged time in mixed-cropped cowpeas.

3.4.1.2 Control

The most endeavoring approach to control the disease is the utilization of host plant resistance. The cowpea germplasm is collected and screened at IITA where two types of resistance have been identified: (1) hypersensitive reactions make cowpea lines functionally immune, and (2) field resistance allows less or null anthracnose development in nurseries. Nature along with inheritance of this resistance is studied at IITA to produce cowpea with varying degrees and high level of stable resistance to anthracnose.

*3.4.2 Cercospora Leaf Spot (*Cercospora canescens *Fellis & Martin and* Cercospora cruenta *Sacc)*

Cercospora leaf spot is a foliar fungal disease that affects a vast number of legumes including cowpea. *Cercospora canescens* and *Cercospora cruenta* (Williams 1975) both cause *Cercospora* leaf spot. They cause severe loss of yield of <40% in cowpea. Although there are not only a variety of resistant lines but also susceptible ones, there is a necessity to identify suitable varieties for cultivation (Booker and Umaharan 2007).

3.4.2.1 Diagnosis and Epidemiology

The initial symptom of *Cercospora* leaf spot in cowpea is the development of tiny, light-colored spot (almost yellow) which later turned to bronze and then dark grayish circular spot. The fungus produces windborne spores in bulk on the abaxial surface of leaf which gives the spots a gray to dark powdery appearance. Symptoms are not usually observed during flowerin4g time. *C. cruenta* occurs in the leaf with more intensity, as it occurs in all seasons when the susceptible lines are planted. Both species are found to be sporulating on pods as well, favored by wet weather (Ratnadass et al. 2012). Yield reductions of cowpea grain attributed by *C. canescens* and *C. cruenta* are about 20% and 40%, respectively, by IITA (1973) (Vaghefi et al. 2018).

3.4.2.2 Control

Crop practices such as intercropping can be applied which includes planting cowpeas in alternate rows along with another suitable nonlegume crop, such as maize, which can limit or eradicate the spread of disease within a field. Chemical approaches include the fungicide's utility to control disease outbreaks when favorable conditions enable disease establishment. The disease develops on older leaves, but early crop survey is difficult to monitor due to complication in distinguishing symptoms from other types of damage. Mancozeb is applied with a maximum of 2–3 applications subsequently after crop flowering and pod development per planting season (Devasirvatham et al. 2012).

3.5 Faba Bean

Faba bean (*Vicia faba* L.) is another important legume seed rich in protein which can adapt to most of the European climatic conditions. Several faba bean cultivars are characterized by varying amount of diets of nutritional value which contain high and/or reduced levels of tannins and a combination of high or low levels of vicine and convicine (VC) (Crépon et al. 2010). This nutritional value was examined in ruminants and monogastric animals. Faba bean has common usage as a staple food in many emerging countries including countries of Asia and Africa (Gago et al. 2014).

3.5.1 *Chocolate Leaf Spot (***Botrytis fabae** *and* **Botrytis cinerea***)*

Chocolate leaf spot of faba bean is caused by *Botrytis fabae* and *B. cinerea*. The disease affects many parts of the world, reducing faba bean yields (Sahile et al. 2008). Serious epidemics were reported in the UK, Tunisia, and Syria (Nene 2003). Fifty percent of faba bean yield loss has been reported in Egypt which is due to chocolate leaf spot and rust diseases, occurring regularly together (Jensen et al. 2010).

3.5.1.1 Diagnosis and Epidemiology

Generally, symptoms include brown-colored spots on the leaves, strips on the stems and petioles, comprehensive darkening of the infected plant, and ultimately death of the infected plant (Motilal and Sreenivasan 2013). The following symptoms are linked to considerable yield losses during extended rainy periods. The age of faba bean influences the severity of chocolate leaf spot (Plantegenest et al. 2007). When observed under artificial conditions, relatively 7-week-old plants had shown more severe disease development than 2-week-old plants. The optimum temperature for infection is around 20 °C and relative humidity is 85% (Nene et al. 2011).

3.5.1.2 Control

The method of breeding disease-resistant cultivars is mostly practiced. Two-cycle procedure has been followed at ICARDA (Nene et al. 2011). In the first cycle, a broad mixture of *B. fabae* isolates with germplasm lines was evaluated, which were collected from leaves of naturally infected plants from the local susceptible cultivars of Syria (Sari et al. 2018). A couple of coalesced-sporulating lesions were developed in the resistant lines, which were detected in the first cycle and then mixed with the isolates collected from such abrasions. Isolates were later eventually inoculated back in the post-screening cycle to the progenies of the resistant lines identified in the first cycle. Subsequently, the outcome of these screenings gave three lines identified as possessing wide-based and stable resistance (Davidson et al. 2016; Sari et al. 2018).

3.5.2 *Rust (*Uromyces viciae-fabae*)*

The rust occurring in most faba bean-growing areas is triggered by *Uromyces viciae-fabae* (syn. *U. fabae*). It is considered to be the most severe constraint of faba bean in Egypt and is conjoint all over the Mediterranean province. Rashid and Bernier (1991) reported faba bean losses of up to 50%.

3.5.2.1 Analysis and Epidemiology

Rust of faba bean is homoecious and two stages are commonly evident: uredial and teleuto. The development of red pustules occurs on either leaves, stems, or petioles, which exhibited small circles. However, the teleutopustules arise on the leaves, and they are commonly present on the stems. They appear to be brown to black. The rust in faba bean crops results in defoliation. The pathogen is also known to infect pea, lentil, and wild-cultured species of *Vicia* and *Lathyrus*. And detailed epidemiological studies are necessary (Eshetu et al. 2018; Hanounik and Hawtin 2011; Zhang et al. 2019).

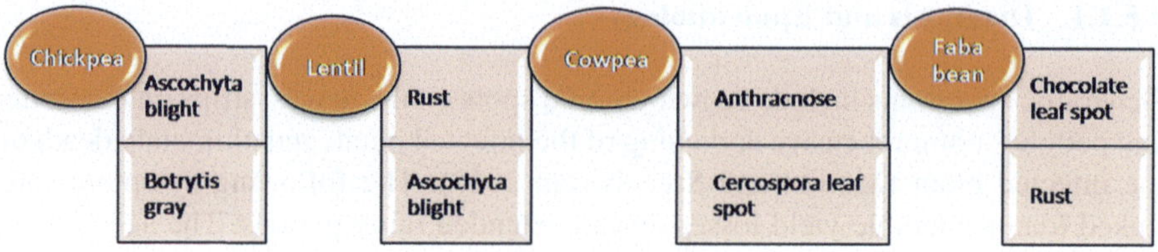

Fig. 3.1 An overview of legumes mentioned in the chapter along with their fungal foliar diseases

Table 3.1 Fungal diseases of legumes and their causal organisms

Sl. no.	Legumes	Fungal diseases	Disease-causing agent	References
1.	Chickpea	*Ascochyta* blight	*Ascochyta rabiei*	Pande et al. (2005)
		Botrytis gray mold	*Botrytis cinerea* Pers. ex Fr.	
2.	Lentil	Rust	*Uromyces viciae-fabae* (Pers.) Schroet	
		Ascochyta blight	*Ascochyta lentis* Bond & Vassil	
3.	Cowpea	Anthracnose	*Colletotrichum lindemuthianum* (Sacc. & Magn.)	
		Cercospora leaf spot	*Cercospora cruenta* (Sacc.)	
4.	Faba bean	Chocolate leaf spot	*Botrytis fabae* and *B. cinerea*	Nene et al. (1988)
		Rust	*Uromyces viciae-fabae*	

3.5.2.2 Control

Practical methods can be applied by utilization of resistant cultivars. There is still an ongoing work at ICARDA and in Canada, where many lines were identified to be resistant. When tested via international nurseries, most of these culture lines were evident with only location-specific resistance. The exceptional case is the resistance of BPL 1179–1 (in Syria, Egypt, and Canada) (Cetin et al. 2002) (Fig. 3.1 and Table 3.1).

3.6 Disease Management of Fungal Foliar Disease

Among the paramount food legumes that are grown globally, the one found in cool season is *Cicer arietinum L.* (chickpea), *Lens culinaris* Medik. (lentil), and *Vicia faba* (faba bean), whereas the one found in warm season is *V. unguiculata* L. (cowpea). Organic pressure markedly minimized the yield of those legumes noticeably. Fungi and viruses are the massive deteriorating factors that affect plants at different growth phases of the legumes (Chen et al. 2006; Ghanem et al. 2015; Walley et al. 2007). Foliar diseases like gray mold and *Ascochyta* blight spawned via varieties of

Botrytis and *Ascochyta* are of vast significance to faba bean, lentil, and chickpeas. In lentil, the genus *Stemphylium* induces foliar disease and in cowpea, *Septoria* species gives rise to leaf spots. Based on published reports, it is found that approximately 45 viruses infect legumes worldwide, but only a few are of economic threat with esteem to certain regions (Gaur et al. 2012; Muehlbauer et al. 2006; Rodda et al. 2017).

In this chapter, a great effort has been made to mark the management of foliar disease of food legumes in both seasons. A successful integrated disease management scheme for economically prime foliar diseases of cowpea, chickpea, faba bean, and lentil has been explored with an allusion to the investigation results on biology, pathogen, and etiology. Integrated disease management strategy (IDM) is the process in which legumes are safeguarded from the yield-reducing consequences of the infectious agent and providing the after commercial insignificance. In this particular system, a discrete constituent of disease controlling plant resistance, backwoods practices, sensible use of fungicides, etc., have to be specific or complementary.

3.6.1 Foliar Disease Management of Food Legumes

Throughout research and development, the prime emphasis to inhibit legume infections is laid upon host resistance and chemical management. The principle of IPM (integrated pest management) has been taken into consideration by IDM (integrated disease management) (Abdullah et al. 2015). The IDM of legumes in a particular area depends upon the genetic resistance, in addition to other components of disease management. Based on the environment, IDM may require a lot of or different components to inhibit foliar diseases (Hema et al. 2014).

In the production of food legumes, the elements of IDM are cataloged in this fashion:

- A host plant resistance
- Disease pressure
- Biotic control
- Agronomic practices

3.6.1.1 Cool Season Legumes

Chickpea The most common foliar diseases in chickpea are *Ascochyta* blight and *Botrytis* gray mold (BGM). This decrepitude was appraised by various workers. Chickpea diseases and their management have been discussed in detail by Varshney et al. (2012). IDM practices are economically vital in potent control of AB (*Ascochyta* blight) and BGM (*Botrytis* gray mold). According to studies in specific areas, several provenances of reluctance to AB were found and the developed genotypes aid to grow the yield during winter in Mediterranean provinces, resulting in the twofold construction potential of chickpeas. And under a high disease pressure, a sufficient level of genetic resistance to BGM is not handy in the cultivated genotype (Tribe et al. 2006). Therefore, the use of handy management options by IDM is vital to mitigate the disease and reduce yield losses.

A union of a fairly resistant type and two chemicals, one during the seedling period and the other at early podding period, issued the best efficient turf control for AB in Syria and Australia (Owati et al. 2017). An IDM package for AB management was initiated by ICARDA in alliance with the Syrian national program. A higher chickpea yield using local variety without other methods was observed with this package. Agronomic and ethnic management of BGM has been exhibited in several countries like India, Bangladesh, and Nepal (Davidson and Kimber 2007; Schreinemachers et al. 2015; Varaprasad et al. 2011; Yadav et al. 2010).

IDM practices for location-specific AB include:

- The seed used that must be free of pathogens
- Treatment of seed with fungicides
- Crop rotation practices
- Deep plowing for burying crowded debris
- Use of disease-resistant genotypes

Lentils The economically vital foliar diseases of lentil are *Ascochyta* blight and rust. *Ascochyta* blight is caused by *A. lentils* producing conidia. It involves the use of resistant cultivators, aiding seed, and seed analysis by foliar spray. It can be maintained by the application of fungicides (Peever et al. 2004). Lentil rust is fostered by *Uromyces viciae-fabae* (Pers.) de Bary, which is an atrocious fungus. The disease arises in the early podding phase as aecia and then into secondary aecis which rapidly shows up a little delay in crop season followed by evaluation of Telia. Integrated management of rust controls volunteer plants in summer and infected lentil debris. It includes the use of clean seeds, suitable fungicide treatment, and host plant resistance. Various rust-resistant cultivators are deployed in different countries, with resistance at CARDIA, Syria, and India (Ammar et al. 2017).

Faba Bean The vital diseases of faba bean are chocolate leaf spot and rust. Another paramount disease of faba bean is brown rust which is spawned by fungus *Uromyces viciae-fabae* Schroet (Mahuku et al. 2016). For controlling the foliar disease of faba bean, the IDM strategy comprises the usage of the disinfected seed, avoiding the spread of disease too quickly, and pursuing crop rotation. In order for the spray program is to be fruitful, regular crop monitoring is crucial. Fungicide application timing depends on the level of disease observed. When high chocolate spot pressure occurs, carbendazim is used, and when rust or *Ascochyta* blight is the problem, then chlorothalonil or mancozeb is used (Varaprasad et al. 2011).

Chocolate spot disease is spawned by *Botrytis fabae*. Initially, chocolate spot occurs on leaves, stem, flowers, etc. as small reddish-brown circular spots. The spot then turned into a gray dead center with a red-brown margin. This disease kills flowers and stems. When the disease spread under favorable conditions, it causes severe defoliation, flower drop, and plant death. The major component of disease management includes resistance because cultural practices and fungicides only give partial crop protection. To take the benefits of high priced fungicides, the faba bean must be grown in early seasons. Chocolate spot control and faba bean yield can be increased by using vinclozolin 50WP, once every 2 weeks. For better management of this disease, different types of fungicides are used such as mancozeb, chlorothalonil, carbendazim, and procymidone (Elliott and Whittington 1979; Noorka and El-Bramawy 2011).

Rust is spawned by *Uromyces viciae-fabae* Pers. Schroet. This rust completes its entire life cycle on faba bean itself. It infects many species. *Uromyces fabae* is short, whitish, and cup-shaped (Barilli et al. 2014).

To reduce the inoculums and avert the disease and future pollution, numerous cultural methods such as suitable plant spacing, appropriate crop rotation, and elimination and burning of crop debris are employed (Sparkes 2016). Field sanitation is vital for reducing losses from faba bean rust. To reduce the chances of primary infection, elimination of infected plant debris and faba bean rotation with nonhost crops play a vital role (Lemke et al. 2007; Rótolo et al. 2015; Wesche et al. 2012). Several control measures are taken to minimize crop losses like the application of mancozeb (0.2%), bayleton (0.05%), and calixin (0.2%) which are fungicides that control pathogenic diseases. The triazole fungicides provide excellent control when applied 72 hours after inoculation. Foliar sprays of mancozeb or chlorothalonil and copper product are valuable in controlling at the time of disease occurrences by a chocolate spot in the same field (Godoy et al. 2016; Hartman et al. 2011) (Fig. 3.2).

3.6.1.2 Warm Season Legume

Cowpea It is the most important legume. *Cercospora* leaf spot, cowpea golden mosaic, and cowpea aphid-borne mosaic are likely of commercial significance. In growing areas of cowpea, *Cercospora* leaf spot is observed. The two most critical diseases in cowpea are cowpea aphid-borne mosaic and cowpea golden mosaic virus. Under field condition, the virus-infected seed gives the basic inoculums, and aphids are accountable for the ancillary extent of the disease. ELISA is one of the important methods for detection of both the seeds and the plant tissue for seed certification project (Nautiyal 2002).

3.7 Sustainable Management of Fungal Foliar Disease

Sustainable management can be defined as a long-term plan of an organized system of plant production practices that will satisfy the present human needs without compromising the economy of future generations and also enhancing environmental

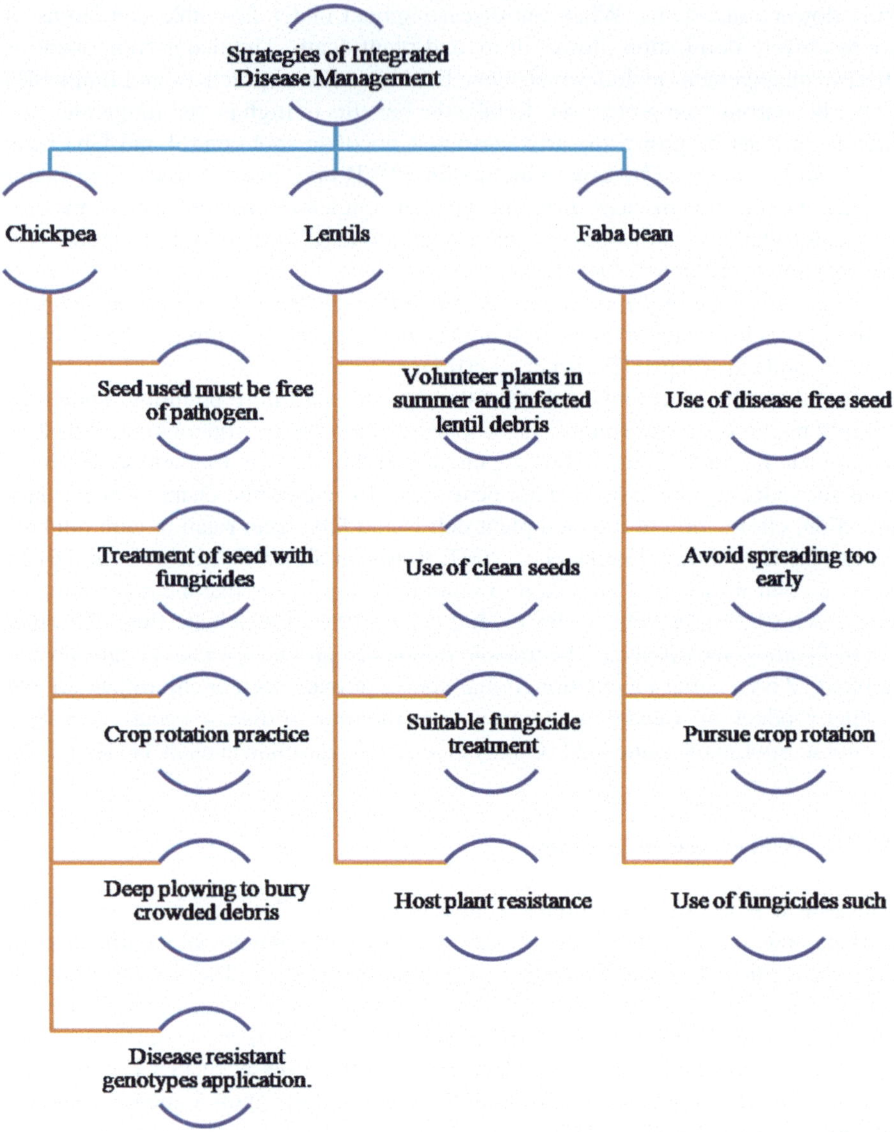

Fig. 3.2 Strategies of integrated disease management in chickpea, lentils, and faba bean

quality. Sustainable agriculture management is carried out for future generations in the form of farming (Folgarait 1998). Sustainable agriculture management comprises the following:

(a) Meet human needs
(b) Natural resources are protected
(c) Prevent degradation of water quality, etc.
(d) Nonrenewable resources efficiently used

(e) Natural organic cycles used
(f) Assure the economic survival of farmers
(g) Institutional incentives created for environmental stewardship

Along with sustainability, new technologies have also improved agricultural production. BMPs are used presently by targeting the applications rather than broadcasting. Cultural practices, biological pest control, new disease resistance hybrids, and many more ways are being implemented (Liang et al. 2016).

3.8 An Outlook for Sustainable Disease Management

Sustainable management of fungal diseases includes exploitation of host resistance, use of synthetic fungicides, and cultural and physical methods, which is discussed below.

The exploitation of host resistance: To control fungal diseases, host resistance is used as an efficient, inexpensive, and effective way. In this segment, available information is integrated for identification of resistance source; molecular markers combine with disease defiance gene identification and improved disease resistance genes (Toyoda et al. 2002). Mainly cultivars are used in host-plant resistance which can tolerate pathogen attack. The interaction between genetic factors in the pathogen and the plant determines the expression of plant resistance. Host-plant resistance could become a deficit when exposed to unsuitable environmental conditions (Andersen et al. 2018). As observed on phoma stem canker (*Leptosphaeria maculans*) of oilseed rape, disease resistance can be dependent on temperature (West et al. 2001) where resistance is expressed at 15 °C but not at 25 °C (Mitrousia et al. 2018).

Protection with fungicides: The usual approach for fungal disease management is the application of fungicides. Disease management in a traditional way is the use of immense spectrum of fungicides as seed treatment chemicals and foliar sprays. Numerous testing were focused on *Cercospora* leaf spot, anthracnose, and powdery mildew, and some trials were on *Macrophomina* blight, web blight, and dry root rot. DMI (demethylation inhibitors) and MBC (methyl benzimidazole carbamate) are the effective fungicides that control foliar diseases. Instantly, after the appearance of disease symptoms, foliar spray was applied followed by second and third sprays after15–20 days from the first spray for anthracnose, powdery mildew, and *Cercospora* leaf spot. Counter to wet and dry root rot seed treatment is applied. Carbendazim is an effective fungicide against dry and wet root rot disease (Rathore et al. 2008; Sumrra et al. 2015). As recommended by the Fungicide Resistance Action Committee (FRAC), various management strategies, markedly, rotation of treatments of a fungicide tank mix of broad spectrum and the fungicides that are selected and integrated fungicide spray program along with elements of disease controlling practices are executed at various levels of organizing bodies of many countries (Vincelli 2002). However, sometimes disease management failures are observed. For example, isolates of *C. kikuchii* (*Cercospora* leaf spot) from soybean fields in the USA were reported to be unaffected by thiophanate-methyl (Soares et al. 2015). Isolates of *Ascochyta* blight of chickpea also reported being unresponsive to chlorothalonil, fluxapyroxad, prothioconazole, and pyraclostrobin. Next-generation fungicides are therefore used which are the derivatives of natural products. These are ecologically safer and effective at reduced doses (Khani et al. 2016; Salam et al. 2011; Pande et al. 2005).

Cultural and physical practices: To terminate seed-borne pathogens, various cultural and physical methods are used to control *Cercospora* foliar blight. In foliar diseases, field cleanliness, crop rotation, etc. is important (Tagne et al. 2008). For example, mung bean seed analysis with gamma rays and storage of 90 days at a subduing effect on root rot fungi (Ikram and Dawar 2017). Computing diversity in the crop rotations maintains the sustainable management of soil-borne diseases. Crop rotation, plant residue management, etc., are productive for controlling diseases in climatic surroundings (Chakraborty 2013; Juroszek and Von Tiedemann 2015).

3.8.1 Challenges for Sustainable Management

Quite a lot of challenges prevail in the enactment of unified supervision, and a lack of suitable understanding of integrated pest management exists among the farmers. For example, gamma rays are used for seed treatment in eliminating the seed-borne pathogen, but in the case of smallholder farmers, it's ineffective because the production of seeds in their farm is done on a small scale. With several studies, disease resistance genotypes were assessed in limited localities or seasons. The pathogen population varies among dissimilar geography, and for that reason, screening of emerging breeding lines for disease resistance should be done in multiple locations (Rebaudo and Dangles 2013;) (Crowder and Harwood 2014).

Various attempts are implemented for the production and application of biopesticides in the undeveloped countries. Several biopesticides just conquer the disease and not effectual as chemical fungicides, and hence the growers are unwilling to use the products. The farmers in those countries are not well equipped with knowledge about the influence of global climate change in disease management which affect the improvement and durability of plant protection chemicals and biocontrol agents which can be a vital task to manage foliar imminent diseases (Afreh-Nuamah and Akotsen-Mensah 2015; Heong et al. 2013).

3.9 Conclusion

Legumes such as chickpea, lentil, faba bean, and cowpea are consumed by the major population worldwide. This chapter dealt with the diagnosis and epidemiology of the fungal foliar diseases such as *Ascochyta* blight, *Botrytis* gray mold, rust, chocolate leaf spot, and *Cercospora* leaf spot and how to control them. In this chapter, the development of the management of foliar diseases of both cold and warm season legumes has been explored. Previous researches were based on resistant sources and chemical control of scarce diseases, whereas the present IDM program lies in identifying, evaluating, merging, and locating distinct components. In spite of the various IDM modules developed to tackle diseases of legumes, but, a gap exists between farmers and scientists. Therefore, IDM technology might be expanded by increasing farmer awareness and the crop residue quality of food legumes which are the vital components of the mixed crop-livestock system.

References

Abdullah ZS, Greenfield BPJ, Ficken KJ, Taylor JWD, Wood M, Butt TM. A new attractant for monitoring western flower thrips, Frankliniella occidentalis in protected crops. Springerplus. 2015;4:1–9.

Adebanjo A, Bankole SA. Evaluation of some fungi and bacteria for biocontrol of anthracnose disease of cowpea. J Basic Microbiol. 2004;44:3–9.

Afreh-Nuamah K, Akotsen-Mensah C. Ghana IPM programme: past, present, and future. Outlooks Pest Manag. 2015;26:4–7.

Ammar MH, Khan AM, Migdadi HM, Abdelkhalek SM, Alghamdi SS. Faba bean drought responsive gene identification and validation. Saudi J Biol Sci. 2017;24:80–9.

Andersen EJ, Ali S, Byamukama E, Yen Y, Nepal MP. Disease resistance mechanisms in plants. Genes (Basel). 2018;339:1–30.

Barilli E, Sillero JC, Prats E, Rubiales D. Resistance to rusts (uromyces pisi and u. viciae-fabae) in pea. Czech J Genet Plant Breed. 2014;50:135–43.

Booker HM, Umaharan P. Identification of resistance to Cercospora leaf spot of cowpea. Eur J Plant Pathol. 2007;118:401–10.

Cernay, Charles, Elise Pelzer, and David Makowski. "A global experimental dataset for assessing grain legume production." Scientific data 3 (2016): 160084.

Cetin L, Dusuneeli F, Albustan S, Mert Z, Wellings C, Torabi M, et al. Virulence of wheat yellow rust in field grown yellow rust differentials, Turkish and regional wheat varieties in Ankara. Meet. Chall. yellow rust Cereal Crop. Proc. First Reg. Conf. Yellow Rust Cent. West Asia North Africa Reg. Karaj, Iran, 8–14 May 2001; 2002;78–89.

Chakraborty S. Migrate or evolve: options for plant pathogens under climate change. Glob Chang Biol. 2013;19:1985–2000.

Chen C, Miller P, Muehlbauer F, Neill K, Wichman D, McPhee K. Winter pea and lentil response to seeding date and micro- and macro-environments. Agron J. 2006;98:1655–63.

Coakley SM, Scherm H, Chakraborty S. Climate change and plant disease management. Annu Rev Phytopathol. 2002;37:399–426.

Crépon K, Marget P, Peyronnet C, Carrouée B, Arese P, Duc G. Nutritional value of faba bean (Vicia faba L.) seeds for feed and food. F Crop Res. 2010;115:329–39.

Crowder DW, Harwood JD. Promoting biological control in a rapidly changing world. Biol Control. 2014;75:1–7.

D'mello, Jp Felix, et al. "Pesticide use and mycotoxin production in *Fusarium* and Aspergillus phytopathogens." European Journal of Plant Pathology 104.8 (1998): 741–751.

Davidson J, Smetham G, Russ MH, McMurray L, Rodda M, Krysinska-Kaczmarek M, et al. Changes in aggressiveness of the Ascochyta lentis population in southern Australia. Front Plant Sci. 2016;7:1–16.

Davidson JA, Kimber RBE. Integrated disease management of ascochyta blight in pulse crops. Ascochyta Blights Grain Legum. 2007;119:99–110.
Devasirvatham V, Tan DKY, Gaur PM, Raju TN, Trethowan RM. High temperature tolerance in chickpea and its implications for plant improvement. Crop Pasture Sci. 2012;63:419–28.
Elliott JEM, Whittington WJ. An assessment of varietal resistance to chocolate spot (Botrytis Fabae) infection of field beans (Vicia faba L.), with some indications of its heritability and mode of inheritance. J Agric Sci. 1979;93:411–7.
Eshetu, Gosaye, Yekedem Bimrew, and Hassen Shifa. "Association of Chocolate Spot and Faba Bean Rust Epidemics with Climate Change Resilient Cultural Practices in Bale Highlands, Ethiopia." Advances in Agriculture 2018 (2018).
Faris MAIE, Takruri HR, Issa AY. Role of lentils (Lens culinaris L.) in human health and nutrition: a review. Med J Nutrition Metab. 2013;6:3–16.
Folgarait PJ. Ant biodiversity and its relationship to ecosystem functioning: a review. Biodivers Conserv. 1998;7:1221–44.
Gago J, Douthe C, Florez-Sarasa I, Escalona JM, Galmes J, Fernie AR, et al. Opportunities for improving leaf water use efficiency under climate change conditions. Plant Sci. 2014;226:108–19.
Gaur PM, Jukanti AK, Varshney RK. Impact of genomic technologies on chickpea breeding strategies. Agronomy. 2012;2:199–221.
Ghanem ME, Marrou H, Biradar C, Sinclair TR. Production potential of Lentil (Lens culinaris Medik.) in East Africa. Agric Syst. 2015;137:24–38.
Girish AG, Rao VP, Moukahal A. Fungi – chickpea. 2019; 1–12. https://sandbox.genebanks.org/resources/general-genebank-management/stog/chickpea94/guidelines93/fungi-for-chickpea/
Godoy CV, Seixas CDS, Soares RM, Marcelino-Guimarães FC, Meyer MC, Costamilan LM. Asian soybean rust in Brazil: past, present, and future. Pesqui Agropecu Bras. 2016;51:407–21.
Hall C, Hillen C, Robinson JG. Composition, nutritional value, and health benefits of pulses. Cereal Chem. 2017;94:11–31.
Hanounik SB, Hawtin GC. Screening for resistance to chocolate spot caused by Botrytis Fabae. Faba Bean Improv. 2011;1982:243–50.
Hartman GL, West ED, Herman TK. Crops that feed the World 2. Soybean-worldwide production, use, and constraints caused by pathogens and pests. Food Secur. 2011;3:5–17.
Hema, Masarapu, et al. "Tropical food legumes: virus diseases of economic importance and their control." Advances in virus research. Vol. 90. Academic Press, 2014. 431–505.
Heong KL, Chien HV, Escalada MM, Trebuil G. Reducing insecticide use in Southeast Asian irrigated rice fields: from experimental ecology to large scale change in practices. Cah Agric. 2013;22:378–84.
Ikram N, Dawar S. Efficacy of wild plant parts in combination with UV irradiation in the control of root rot fungi. Walailak J Sci Technol. 2017;14:225–34.
Isman MB. Plant essential oils for pest and disease management. Crop Prot. 2000;19:603–8.
Jensen ES, Peoples MB, Hauggaard-Nielsen H. Faba bean in cropping systems. F Crop Res. 2010;115:203–16.
Jones JT, Haegeman A, Danchin EGJ, Gaur HS, Helder J, Jones MGK, et al. Top 10 plant-parasitic nematodes in molecular plant pathology. Mol Plant Pathol. 2013;14:946–61.
Juroszek P, Von Tiedemann A. Linking plant disease models to climate change scenarios to project future risks of crop diseases: a review. J Plant Dis Prot. 2015;122:3–15.
Kaiser, W. J., et al. "Foliar diseases of cool season food legumes and their control." Linking research and marketing opportunities for pulses in the 21st century. Springer, Dordrecht, 2000. 437–455.
Kaiser WJ, Ramsey MD, Makkouk KM, Bretag TW, Açikgöz N, Kumar J, et al. Foliar diseases of cool season food legumes and their control. Cur Plant Sci Biotech Agri. 2011:437–55.
Khani M, Davidson JA, Sosnowski MR, Scott ES. Survival, transmission and control of Phoma koolunga in field pea seed and reaction of field pea genotypes to the pathogen. Australas Plant Pathol. 2016;45:91–102.
Lemke RL, Zhong Z, Campbell CA, Zentner R. Can pulse crops play a role in mitigating greenhouse gases from North American agriculture? Agron J. 2007;99:1719–25.

Liang J, Crowther TW, Picard N, Wiser S, Zhou M, Alberti G, et al. Positive biodiversity-productivity relationship predominant in global forests. Science. 2016;354(6309).

Mahuku G, Chen J, Shrestha R, Narro LA, Guerrero KVO, Arcos AL, et al. Combined linkage and association mapping identifies a major QTL (qRtsc8-1), conferring tar spot complex resistance in maize. Theor Appl Genet. 2016;129:1217–29.

Mitrousia GK, Huang YJ, Qi A, Sidique SNM, Fitt BDL. Effectiveness of Rlm7 resistance against Leptosphaeria maculans (phoma stem canker) in UK winter oilseed rape cultivars. Plant Pathol. 2018;67:1339–53.

Motilal LA, Sreenivasan TN. Birth of trinitario cacao: history intertwined with myths and edaphic and climatic factors. Combat Clim Chang Agric Perspect. 2013:305–24.

Muehlbauer FJ, Cho S, Sarker A, McPhee KE, Coyne CJ, Rajesh PN, et al. Application of biotechnology in breeding lentil for resistance to biotic and abiotic stress. Euphytica. 2006;147:149–65.

Nautiyal P. Post-harvest operations. INPhO – Post-harvest compend. GROUNDNUT. 2002; 3–126.

Negussie T, Pretorius ZA. Lentil rust: present status and future prospects. Crop Prot. 2012;32:119–28.

Nel B, Steinberg C, Labuschagne N, Viljoen A. Evaluation of fungicides and sterilants for potential application in the management of *Fusarium* wilt of banana. Crop Prot. 2007;26:697–705.

Nene YL. Multiple-disease resistance in grain legumes. Annual review of phytopathology. 1988 Sep;26(1):203–17.

Nene YL. Multiple-disease resistance in grain legumes. Annu Rev Phytopathol. 2003;26:203–17.

Nene YL, Hanounik SB, Qureshi SH, Sen B. Fungal and bacterial foliar diseases of pea, lentil, faba bean and chickpea. Cur Plant Sci Biotech Agri. 2011:577–89.

Noorka IR, El-Bramawy MAS. Inheritance assessment of chocolate spot and rust disease tolerance in mature faba bean (*Vicia faba* L.) plants. Pakistan J Bot. 2011;43:1389–402.

Owati AS, Agindotan B, Pasche JS, Burrows M. The detection and characterization of QoI-resistant didymella rabiei causing ascochyta blight of chickpea in Montana. Front Plant Sci. 2017;8:1–11.

Pande S, Galloway J, Gaur PM, Siddique KH, Tripathi HS, Taylor P, et al. Botrytis grey mould of chickpea: a review of biology, epidemiology, and disease management. Aust J Agric Res. 2006;57:1137–50.

Pande S, Siddique KHM, Kishore GK, Bayaa B, Gaur PM, Gowda CLL, et al. Ascochyta blight of chickpea (*Cicer arietinum* L.): a review of biology, pathogenicity, and disease management. Aust J Agric Res. 2005;56:317–32.

Parsa S, Morse S, Bonifacio A, Chancellor TCB, Condori B, Crespo-Pérez V, et al. Obstacles to integrated pest management adoption in developing countries. Proc Natl Acad Sci. 2014;111:3889–94.

Pedersen EA, Morrall RAA, Mc Cartney HA, BDL F. Dispersal of conidia of *Ascochyta fabae* f. sp. lentis from infected lentil plants by simulated wind and rain. Plant Pathol. 1994;43:50–5.

Peever TL, Salimath SS, Su G, Kaiser WJ, Muehlbauer FJ. Historical and contemporary multi-locus population structure of *Ascochyta rabiei* (teleomorph: *Didymella rabiei*) in the Pacific Northwest of the United States. Mol Ecol. 2004;13:291–309.

Peshin R, Bandral RS, Zhang WJ, Wilson L, Dhawan AK. Integrated pest management: a global overview of history, programs, and adoption. Integr Pest Manag. 2009;1:1–49.

Plantegenest M, Le May C, Fabre F. Landscape epidemiology of plant diseases. J R Soc Interface. 2007;4:963–72.

Raseduzzaman M, Jensen ES. Does intercropping enhance yield stability in arable crop production? A meta-analysis. Eur J Agron. 2017;91:25–33.

Rashid MH, Hossain MA, Kashem MA, Kumar S, Rafii MY, Latif MA. Efficacy of combined formulations of fungicides with different modes of action in controlling Botrytis Gray Mold disease in chickpea. Sci World J. 2014;2014:639246.

Rashid KY, Bernier CC. The effect of rust on yield of faba bean cultivars and slow-rusting populations. Canadian Journal of Plant Science. 1991;71(4):967–72.

Rathore HS, Ishratullah K, Varshney C, Varshney G, Mojumdar SC. Fungicidal and bactericidal activity of metal diethyldithiocarbamate fungicides: synthesis and characterization. J Therm Anal Calorim. 2008;94:75–81.

Ratnadass A, Fernandes P, Avelino J, Habib R. Plant species diversity for sustainable management of crop pests and diseases in agroecosystems: a review. Agron Sustain Dev. 2012;32:273–303.

Rebaudo F, Dangles O. An agent-based modeling framework for integrated pest management dissemination programs. Environ Model Softw. 2013;45:141–9.

Rodda MS, Davidson J, Javid M, Sudheesh S, Blake S, Forster JW, et al. Molecular breeding for Ascochyta blight resistance in lentil: current progress and future directions. Front Plant Sci. 2017;8:1–11.

Rótolo GC, Montico S, Francis CA, Ulgiati S. How to land allocation and technology innovation affect the sustainability of agriculture in Argentina Pampas: an expanded life cycle analysis. Agric Syst. 2015;141:79–93.

Sahile S, Ahmed S, Fininsa C, Abang MM, Sakhuja PK. Survey of chocolate spot (*Botrytis fabae*) disease of faba bean (*Vicia faba* L.) and assessment of factors influencing disease epidemics in northern Ethiopia. Crop Prot. 2008;27:1457–63.

Salam MU, Davidson JA, Thomas GJ, Ford R, Jones RAC, Lindbeck KD, et al. Advances in winter pulse pathology research in Australia. Australas Plant Pathol. 2011;40:549–67.

Sari E, Bhadauria V, Ramsay L, Hossein Borhan M, Lichtenzveig J, Bett KE, et al. Defense responses of lentil (Lens culinaris) genotypes carrying non-allelic ascochyta blight resistance genes to Ascochyta lentis infection. PLoS One. 2018;13:1–27.

Schreinemachers P, Balasubramaniam S, Boopathi NM, Ha CV, Kenyon L, Praneetvatakul S, et al. Farmers' perceptions and management of plant viruses in vegetables and legumes in tropical and subtropical Asia. Crop Prot. 2015;75:115–23.

Soares APG, Guillin EA, Borges LL, Da Silva ACT, De Almeida ÁMR, Grijalba PE, et al. More Cercospora species infect soybeans across the Americas than meets the eye. PLoS One. 2015;10:1–20.

Sparkes D. Field crops. Encycl Appl Plant Sci. 2016:18–22.

Sprent JI, Gehlot HS. Nodulated legumes in arid and semi-arid environments: are they important? Plant Ecol Divers. 2010;3:211–9.

Sumrra SH, Hanif M, Chohan ZH. Design, synthesis and in vitro bactericidal/fungicidal screening of some vanadyl(IV)complexes with mono- and di-substituted ONS donor triazoles. J Enzyme Inhib Med Chem. 2015;30:800–8.

Tagne A, Feujio TP, Sonna C. O.42-Essential oil and plant extracts as potential substitutes to synthetic fungicides in the control of fungi. Divers Crop Prot. 2008:12–5.

Toyoda K, Collins NC, Takahashi A, Shirasu K. Resistance and susceptibility of plants to fungal pathogens. Transgenic Res. 2002;11:567–82.

Tribe HT, Baker KF, Cook RJ. Biological control of plant pathogens. J Appl Ecol. 2006;12

Vaghefi N, Kikkert JR, Hay FS, Carver GD, Koenick LB, Bolton MD, et al. Cryptic diversity, pathogenicity, and evolutionary species boundaries in Cercospora populations associated with Cercospora leaf spot of Beta vulgaris. Fungal Biol. 2018;122:264–82.

Vandana UK, Chopra A, Bhattacharjee S, Mazumder PB. Microbial biofertilizer: a potential tool for sustainable agriculture. Microorg Green Rev. 2017:25–52.

Varaprasad KS, Sharma SK, Sivaraj N, Sarker A. Integrated gene resource management of under-utilized legumes in India. Euphytica. 2011;180:49–56.

Varshney RK, Kudapa H, Roorkiwal M, Thudi M, Pandey MK, Saxena RK, et al. Advances in genetics and molecular breeding of three legume crops of semi-arid tropics using next-generation sequencing and high-throughput genotyping technologies. J Biosci. 2012;37:811–20.

Vincelli P. QoI (strobilurin) fungicides: benefits and risks. Plant Heal Instr. 2002;

Walley FL, Clayton GW, Miller PR, Carr PM, Lafond GP. Nitrogen economy of pulse crop production in the Northern Great Plains. Agron J. 2007;99:1710–8.

Wesche K, Krause B, Culmsee H, Leuschner C. Fifty years of change in Central European grassland vegetation: large losses in species richness and animal-pollinated plants. Biol Conserv. 2012;150:76–85.

West JS, Kharbanda PD, Barbetti MJ, Fitt BDL. Epidemiology and management of *Leptosphaeria maculans* (phoma stem canker) on oilseed rape in Australia, Canada and Europe. Plant Pathol. 2001;50:10–27.

Williams RJ. Diseases of cowpea (*Vigna unguiculata* (L) Walp.) in Nigeria. PANS Pest Artic News Summ. 1975;21:253–67.

Yadav SS, Redden R, McNeil DL, Patil SA. Climate change and management of cool season grain legume crops. Clim Chang Manag Cool Seas Grain Legum Crop. 2010; 1–460.

Zhang C, Dong Y, Tang L, Zheng Y, Makowski D, Yu Y, et al. Intercropping cereals with faba bean reduces plant disease incidence regardless of fertilizer input; a meta-analysis. Eur J Plant Pathol. 2019:1–12.

Chapter 4
Omics Approaches in Chickpea *Fusarium* Wilt Disease Management

Abeer Hashem, Baby Tabassum, and Elsayed Fathi Abd_Allah

4.1 Introduction

Chickpea is an important founder of crops in agriculture, having diploid ($2n = 16$) chromosome number. It belongs to legumes and papilionoid (subfamily) from its wild *Cajanus reticulatus* ancestor present in Turkish Kurdistan dating back (8000–9000) years (Lev-Yadun et al. 2000) and considered a major source of human food due to the presence of lysine-rich protein. It is an important legume and pulse crop in the world having 41–50.8% carbohydrates, 3–6% oil, 17–24% protein, and considerable amount of other minerals like phosphorus, magnesium, calcium, potassium, iron, zinc, and manganese. Chickpea also plays an important role as an alternate rotation crop followed by cereals and manages soil fertility and productivity by improving the N fertilization (nitrogen-fixing ability) from the atmosphere (Jiménez Díaz et al. 2015). Over the past few years, it is stated that chickpea productivity has been marginal decreases due to the effect of biotic factors (*Fusarium* wilt and pod borer) and abiotic factors. Reducing the pressure of these factors (biotic and abiotic) is important to increase production. Chickpea ranked second among the

A. Hashem
Botany and Microbiology Department, College of Science, King Saud University, Riyadh, Saudi Arabia

Mycology and Plant Disease Survey Department, Plant Pathology Research Institute, ARC, Giza, Egypt

B. Tabassum
Toxicology Laboratory, Department of Zoology, Government Raza Post Graduate College Rampur, Rampur, Uttar Pradesh, India

E. F. Abd_Allah (✉)
Plant Production Department, College of Food and Agricultural Sciences, King Saud University, Riyadh, Saudi Arabia
e-mail: eabdallah@ksu.edu.sa

B. P. Singh et al. (eds.), *Management of Fungal Pathogens in Pulses*, Fungal Biology, https://doi.org/10.1007/978-3-030-35947-8_4

important food legume crops in tropical, subtropical, South, and West Asia. Overall, about 1.35×10^7 ha of chickpea are growing and yield about 1.31×10^7 in more than 50 countries. Chickpea is used not as a valuable crop for export in developed countries but a good source of protein supplement in cereal-based diets in developing countries. Chickpea is generally grown under the rainfed condition and depends on available soil water showing drought tolerance over the year. *Fusarium oxysporum* f. sp. *ciceris* (FOC) affects the chickpea crop by inducing wilt disease, more damaging worldwide for their occurrence, and accounts 90% annual yield losses worldwide. The disease was first reported by Butler in 1918, but etiology was not confirmed until 1940 and later was spread in Americas, Europe, and Africa but not reported in Australia. *Fusarium* wilt has become a limiting factor for chickpea production in the Mediterranean basin, the Indian subcontinent, and America. The most important symptoms of wilt, i.e., the patch in group form and occurs at any stage and spread across a field (Haware 1990). The main reason for *Fusarium* wilt is soilborne pathogen and observing signs like delaying crown, leaf anomalies, and rolled brown leaves. The number of strains is unknown to the soilborne pathogen and is difficult to control without solid information and identification of the pathogen (Cha et al. 2016).

The susceptible varieties showed symptoms in 25 days after sowing such as including flabbiness in leaves tailed by a dull green streak, dehydration, and downfall of the plant. Though disease marks are commonly more visible at the initiation of flowering for 6–8 weeks, in some studies, it is reported that it appeared at the podding stage. The leaves dropping has occurred in the upper part of the plant, but within a few days, it ensures on the whole plant. In partial wilt, few branches were affected initially, but later roots of affected material affect the nearby plants. In partial wilting, no color discoloration was recorded visually. In general, symptoms of the disease occur at any stage of plant growth (Jiménez Díaz et al. 2015) while more visible at the early stage of flowering and appears at the podding stage (late wilt). Late wilted plants exhibited falling of petioles, rachis, and leaflets as well as necrosis and discoloration of foliage (Jiménez Díaz et al. 2015). Early *Fusarium* wilt affects more than late wilting. However, late wilted plants produce lighter, rougher, and duller seeds as compared to normal (Haware and Nene 1980; Navas-Cortés et al. 2000). If the cross-sectional study was done on the affected plant, a dark brown color discoloration was observed in xylem tissues. The discoloration was also recorded in vascular tissues of roots as well as in stems. The symptoms were also recorded as cavity formation among xylem and phloem, medulla and cortical parenchyma, and cell proliferation in vascular cambium.

During the defense mechanism, the plant uses many molecular signals or protein receptors to know the presence of microbes. Two modes of pathogen recognition used by the host, i.e., effector-triggered immunity (ETI) and pathogen-triggered immunity (PTI). The invariant epitope types are called microbe-associated molecular patterns (MAMPs) and are composed of flagellin, chitin, and lipopolysaccharides that help spread the disease. Moreover, pathogen-induced danger-associated molecular patterns are composed of fructans, callose, and glucans. As a result, host secretes effector R protein domains have nibblers act as PTI. Studies also reported that the sensing of bacteria produce siderophores and fungi serve as MAMPs and hydroxyproline and rapid alkalization factors, but their role was not clear yet in defense mechanism. The current has described the chickpea *Fusarium* wilt etiology, occurrence, and management practices including the most recent molecular breeding, high-throughput sequencing techniques, as well as identification of transcription factors that could favor the crop and enhance the tolerance mechanism to control the disease.

4.2 Casual Organism and Symptoms

It is caused by *Fusarium oxysporum* f. sp. *ciceris* [*Fusarium oxysporum* Schlecht. f. sp. *concerns* (Padw.) Matuo & Sato] (Jimenez-Fernandez et al. 2011; Haware 1990). The aerial mycelium in the first appearance was whitish and cotton, on potato sucrose agar, potato dextrose agar and under UV light, but turn into salmon in color and some cases, remain white (Jimenez Diaz et al. 2011). *Fusarium* wilt of chickpea produces microconidia, macroconidia, and chlamydospores. The microconidia are elliptical or tubular and straight. Macroconidia are thinner than microconidia and typically 3–5 septate or fusoid, while chlamydospores are produced in 15-day-old cultures and infected chickpea tissues, smooth or rough-walled (Castro et al. 2012; Jimenez Diaz et al. 2011). Maximum sporulation was recorded at pH ranges from 7.1 to 7.9 (Jimenez Diaz et al. 2011). Hyphae are septate and split abundantly. Optimum growth was recorded at 25–27 °C and pH 5.1–5.9 and liable on strains.

4.3 Epidemiology

The severity of the chickpea wilt is depending upon the pathogen, genotypes, pathogenic races, inoculum density, environmental condition, and cultivar sensitivity. The activity of the wilting disease was triggered by a combination of pathogen activities. It includes fungus mycelium in the xylem that produced contaminant components that affect host defense response, production of gels, teloses, and vessel crushing by the propagation of linked parenchyma cells (Beckman 1987). The mycelium might survive as a pathogen in seed, soil and toxic residues (crop), roots, and stem tissue concealed in the soil for more than 6 years or even in absence of host (Singh et al. 2008). Dicotyledonous weeds that don't show the symptoms but have the infection that could enhance the pathogen activity and survived in fallow soils. Moreover, infected soil is an important source of primary inoculum for the development of *Fusarium* wilt (Al-taae et al. 2013). The transmission can also be done by the seed and can survive in plant debris as well as in the soil. Moreover, it also observed that fungus chlamydospore was present in soil freely (Haware et al. 1996), seed hilum (Haware et al. 1978), and cotyledon axis (Shakir and Mirza 1994). Chlamydospores or mycelia are the main and basic sources of infection, even the conidia of the fungus are short-lived, while chlamydospores can remain feasible up to the next available crop in the field (Chand and Khirbat 2009). Chlamydospore production is contingent on the nutrient availability of the inoculum. Fungal inoculum may be exposed to lower nutrient levels in the field condition as compared to grow under well-fed macroconidia form under agar media (Schippers and Van Eck 1981). The pathogen grows very well in roots and stems in apparently looking good condition but concealing adequate fungus (Trapero Casas and Jimenez Diaz 1985). The pathogen remains dormant until triggered to germinate when carbohydrate is released from decaying tissue or roots, present in the form of chlamydospores (Schippers and Van Eck 1981). The provocation for germination could be the host or non-host plant roots or plant wreckage (Nelson 2012), after the germination of chlamydospores, conidia, hyphae, and new chlamydospores is formed. After conidia and hyphae production, thallus formation took place and leads to chlamydospore production in 2–3 days if suitable condition prevailed (Beckman and Roberts 1995). By penetration of the epidermal cells, attack on the roots occurs on the host or non-host plants (Beckman and Roberts 1995) and caused vascular disease (Stover 1970). The infiltration occurs directly or by wounds (Nelson 2012), the common sites for infiltration are the root tip of both tap and lateral roots (Lucas 1998). The infiltration is stopped by different factors, such as fungal compounds, and inhibits the spore formation (fungal), plant surface structures, and germ tube production (Mendgen et al. 1996). The more adverse form is, mycelium moved through intercellular root cortex and finally reaches to xylem vessels during colonization and remains within the xylem vessels and colonize in the host (Bishopt and Cooper 1983).

4.4 Breeding

The *Fusarium* wilt activity can be reduced in the host using breeding approaches in chickpea crop. Breeding approaches involved availability of genetic diversity considered the most important step for a breeding program, wild relatives, and selection of desirable plant for trait and disease resistance and evaluate the plant for commercial production (Salimath et al. 2007). As chickpea is a self-pollinated crop, it requires genes to fix the breeding problems by pure lines development. Initial screening was done by mass or pure line selection and later crossing programs and alteration in pedigree and bulk methods were employed for segregating generation (Gaur et al. 2012; Millan et al. 2015). In the intraspecific hybrid program, the single cross method was used in desi and Kabuli chickpea genotypes with variant genetic history (Berrada et al. 2007). Parents from desi varieties have been used for gene transfer in Kabuli varieties against *Fusarium* wilt resistance, as parents from Kabuli parents are used to improve large size seed and seed quality in desi variety (Gaur et al. 2007). The breeding development efforts were also made for interspecific crosses and enhance genetic diversity and interrogate useful genes from wild cicer into cultivated spp. The FOC resistance has been recognized from desi germplasm as well as in wild *Cicer* spp. (Kaiser et al. 1994). For genetic gains enhancement, there is a need precise and efficient selection of segregating populations (Gaur et al. 2012). For successful wilt, sick breeding programs hot spot location, field, greenhouse and laboratory methods have been used for the selection of resistance varieties (Gaur et al. 2007). It has been reported about 5174 Kabuli genotypes were screened against *Fusarium* wilt resistance at ICARDA, and about 110 genotypes were recognized as resistant. *Fusarium* wilt resistance depends upon monogenic or oligogenic depending upon the resistance resource (Sharma and Muehlbauer 2007; Upadhyaya et al. 1983; Sharma et al. 2005). It is also reported that FOC genetic resistance cultivar contains three independent genes (h1, h2, and h3) (Singh et al. 2014). Moreover, it is also suggested that late wilting was controlled by the presence of any one gene nut combination of two genes confirm the wilt resistance in chickpea (Castro et al. 2012; Jiménez Díaz et al. 2015). Similar results also stated that resistance was confirmed by the presence of these genes in the combine or individual form (Tullu et al. 1999). Some ICARDA lines, i.e., WR-315, CA-1938, and CA2139, contain these genes (Halila et al. 2009; Rubio et al. 2003). However, the genetic of resistance for some chickpea races like 1B/C and 6 is still unknown.

4.5 Genetic and Pathogen

The first name of the fungus was *Fusarium orthoceras* apple and swollen. var. cicerone by Padwick and modified by Chattopadhyay and Sen Gupta and was renamed as *F. oxysporum* Schl. f. sp. *ciceri* (Padwick) Snyder and Hansen. *Fusarium oxysporum* is among the monophyletic origin in the *Fusarium oxysporum* complex of the gibberella clade and considered as polyphyletic and currently known as *Fusarium oxysporum* (Schlechtend. Fr.*)* f. sp. *ciceris* (Padwick) Matuo & K. Sato. *Fusarium* is the only pathogen in *Cicer* sp. (Kaiser et al. 1994), and *oxysporum* is an attack on root tissue in faba bean, lentil, and pea and recorded as symptomless carters for the pathogen (Trapero Casas and Jimenez Diaz 1985).Yellow or wilting syndromes along with brown discoloration were recognized based on two pathotypes and induce in sensitive chickpeas. The recorded symptoms are considered slow, foliar yellowing and death of plant at a later stage while wilting is considered reckless, adverse chlorosis, flabbiness, and plant death during an early stage of growth (Trapero Casas and Jimenez Diaz 1985).The susceptibility of the pathogen depends upon the races and efficient use of available resources for the chickpea breeding program. The identification of the races against pathogens is simple but depends upon the cost, available resources, and facilities. So, there is a dire need to develop new methods that are more rapid and effective, and reproducible identification of pathogen and races is used to determine the diversity and resistance among the genotypes. Polymerase chain reaction (PCR)-based molecular markers have been used to determine the *Fusarium oxysporum* f. sp. *ciceris* and its related pathogen races identified by the method developed by Jiménez-Gasco et al. (2001).

The screening and legacy of the gene of interest (GOI) and traits are possible now with the development of marker-assisted selection (MAS) and provide beneficial information to exploit the genes useful for agronomic traits (Allahverdipoor et al. 2011). Molecular markers are an important tool for identification, characterizing, and screening and determine the diversity among the pathogens and diseases. Commonly, internal transcribed spacer (ITS) markers are used for classification and screening of the fungi (White et al. 1990), while ITS data is not enough for complex identification and diverse gene information; therefore, it is not suitable for genetic diversity or characterization of fungus. The *Fusarium* genus is improbable as compared to the genetic study of *F. oxysporum* f. sp. *fragariae has not yet been reported.* Among the various available technologies, restriction fragment length polymorphism (RFLP) markers are important rDNA used to determine the genetic diversity of plant pathogenic fungi. It is also used to group the isolated strains with low cost (Kachuei et al. 2015). Based on symptoms, the two pathogens were genetically distinguished by random amplified polymorphic DNA markers (RAPD) and sequence characterized amplified region (SCAR). The specific *Fusarium* assays were successfully characterized using RAPD and SCAR molecular markers. Another study stated that evaluation and screening of resistant wilt lines were done against *Fusarium* by using RAPD and SSR molecular markers. The results represent that about 70% cultivars were resistant to disease while 30% showed susceptibility for wilt response. SSR marker (TA194) recorded an 85% probability locus at wilt resistant among the total primer used, and it was later reconfirmed by the receiver operating characteristic curve (Ahmad et al. 2014). Gowda et al. (2009) former designing the linkage map for *FOC* 1–5 gene resistance races with SSR and RAPD in recombinant inbred lines (RILS) developed by sensitive and resistant parents. About eight races were recognized as the specific fungus, out of which six are more infectious (Jimenez-Diaz et al. 1993). Introgression of *Ascochyta* blight resistance with double podding traits in chickpea was confirmed by marker-assisted backcrossing. SSR markers are used in separate backcross generation to assist in selection against the resistance of *Fusarium* (Varshney et al. 2014). SCAR markers are used for *Ascochyta* blight resistance to determine the QTLs in chickpea, and respected QTLs were identified, i.e., SCY17590 and SCAE19336, tightly linked with *Ascochyta* blight resistance gene at QTLAR2 location (Iruela et al. 2006) and later on successfully used for tagging in chickpea resistance lines for germplasm collection (Imtiaz et al. 2008). Combinations of SCAR with a codominant marker (CaETR) linked with QTLAR1 for *Ascochyta* tagging and help to identify the resistance alleles from a core collection of resistant cultivars (Madrid et al. 2013). Near-isogenic lines (NILs) were developed by using STMS markers that are tightly linked FOC 5 and FOC 01 for the selection of susceptible genotypes and resistant genotypes in LG2 and LG5 (Castro et al. 2010). Moreover, NILs are used as a valuable tool for mapping, refining the target region and selection of the desired gene for resistance to foc0 (Jendoubi et al. 2016). Jendoubi et al. (2016) reported that the results obtained from the population were useful for position refining of the target area involved in resistance mechanisms. Similar results were obtained by Ali et al. (2015) that identify the target regions associated with growth habit and double-podding base morphological position-based markers that are used in chickpea.

4.6 Integrated Genomic Approaches

The identification and construction of the genetic map of the segregating population is the foremost objective of the breeders. Efforts have been made to construct the genetic map using molecular markers for tagging traits and site-specific gene of interest in chickpea (Millan et al. 2010; Millan et al. 2015). The first maps were constructed using the isozymes F2 population from interspecific crosses (Gaur and Slinkard 1990). Many researchers reported identified genes regarding flower color, wilt resistance (*Fusarium*), double pod, and growth habit (Gaur and Slinkard 1990; Kazan et al. 1993; Cobos et al. 2005), and other agronomic characters and *Ascochyta* blight resistance linked QTLs were identified on these maps (Lichtenzveig et al. 2006). The larger numbers of maps were derived from crosses with *C. reticulatum* as well as many markers identification related to specific traits. However, the populations derived from interspecific crosses were made due to microsatellite markers and exploit more genetic polymorphisms among the chickpea genotypes (Cobos et al. 2007). The first transcriptome study for the chickpea genome was done with the advancement of next-generation sequencing (Hiremath et al. 2011). With the advancement of transcriptome information, detail genetic maps were made using large-scale molecular markers (Hiremath et al. 2012; Thudi et al. 2011). The availability of the draft genome sequencing in desi and Kabuli varieties would also facilitate the genetic population used for mapping and positioning of the QTLs in chickpea genome (Ali et al. 2015). Omics approaches gathered genomic information and triggered molecular markers development of tightly linked QTLs (Kumar et al. 2011).

4.7 Transcription Factors

Recent advances in molecular plant sciences boost the knowledge, and transcriptomic emerged as a powerful method to understand differential genic response over specific time-bound fashion. Transcriptomic is the techniques used to study the whole set of RNA transcripts (coding and non-coding) of a cell at a specific time and conditions. Expression analysis of tissue under different growth conditions reveals the regulatory network of the responsive gene for that specific stage or conditions, it could also help to annotate those genes which were previously unannotated due to lack of information. TF has the function to regulate the cell development, differentiation, and growth by tagging specific site with DNA or multiple sites and triggered the activation or repression of the TF through various mechanism and interaction, i.e., DNA-protein, protein-protein, and alteration in chromatin structure (Kusuya et al. 2018). The soilborne fungus is a causal agent of chickpea wilt disease. The infection includes root identification, colonization, penetration, adhesion, and penetration of the root cortex, and hyphal proliferation within the xylem vessels are controlled by transcription factors (TFs). Transcriptome analysis based on RFLP and RAPD-based cDNA techniques were used and identified many defense-related genes in chickpea (Gurjar et al. 2012). Moreover, next-generation sequencing identified microRNA responsive genes regulating plant development and pathogen growth depending on target genes (Kohli et al. 2014). *Fusarium* spp. produced about 50 unique types of secondary metabolites, i.e., growth regulators, pigments, and mycotoxins, that are important for feed and food concerns. TFs have been shown to manage the mycotoxin biosynthesis compound that is favorable for other pathogenic *Fusarium* species (Brown et al. 2014).

Identification of *FolCZF1* encoded for (C2H2) transcription factor. It is also known to affect pathogenicity in wheat (*F. graminearum*) and rice (*Magnaporthe oryzae*). The critical role of gene *FolCZF1* is to produce fusaric acid and regulate the expression of fusaric acid biosynthesis. Fusaric acid (FA) taking part in the severity of *Fusarium* diseases, i.e., damping off, vascular wilt, and root rot (Ding et al. 2015). Fusaric acid is linked with vascular wilt symptoms caused by *F. oxysporum*; some transcription factors are involved in the regulation of virulence and FA biosynthesis. *FolCZF1* affects the FA and influence the virulence (Yun et al. 2019). Moreover, *FolCZF1* is also reported that it requires secondary metabolism and early host infection (Yun et al. 2019). Zinc finger proteins (C_2H_2) are widely studied in filamentous fungi.

A similar study was conducted to determine the molecular basis of wilt disease in chickpea by comparing the analysis of the transcriptome of resistant and susceptible wilt cultivars under *Fusarium oxysporum* f. sp. *ciceri* and controlled condition. Analysis results stated that novel genes with differential or unique expression causative to lignification, hormonal balance, plant defense signaling, and ROS. Moreover, the study also provides information about the functional characterization of the genes involved in resistance mechanism and their use in a breeding program against wilt resistance and tolerance mechanism as well as target pathogen identification for the facilitation of the development of novel control management strategy (Upasani et al. 2017). Microscopic, proteomic, and metabolic approaches are also used to characterize the chickpea cultivars under *Fusarium oxysporum* interaction. The resulting expression at the microscopic level stated that differential colonization of FOC was present in susceptible and resistant genotypes. It is also reported that resistant host severely restricted the pathogen growth while opposite results were observed in susceptible cultivars. Moreover, proteomics and metabolomics results notified that the upregulation of several metabolic pathways was observed in resistant genotypes (Kumar et al. 2015; Kumar et al. 2016; Upasani et al. 2016).

ROS played an important role in recognized insight and defense signaling, but their redox relation in plant is still unknown for the defensive network. A study was conducted to determine the role of FOC 1 by inducing redox-responsive transcript for regulating defense signaling in chickpea. Microscopic studies emphasized invasion and colonization along with tissue damage and confession of degraded products at the xylem vessels in diseased roots area. Due to confession clogging of the xylem vessels incompatible hosts while resistant plant not. Assays related to lipid peroxidation represent membrane injury, and other remarkable changes were recorded such as cell shrinkage and gradual nuclear depression in fungal ingress. Moreover, qPCR results showed expression of redox regulators, cellular transport, and transcription factors in FOC 1 analysis. Functional analysis results stated that respiratory homolog, vacuolar sorting receptors, and zinc finger domain TF provide deep insight regarding the complex structure of wilt disease defense mechanism in chickpea as well as other legume crops (Gupta et al. 2013). The study also reported that chickpea transcript is used for involvement to regulate the redox state when infection occurs due to FOC 1 races (Gupta et al. 2009; Ashraf et al. 2009; Gupta et al. 2010; Garcia-Limones et al. 2002). Moreover, it is also reported that modification in the RBOH recorded regulatory role during an invasion in resistant plants while sensitive plants do not show similar variation. The other modification in OCP and FSD has a role in ROS signaling and OCP considered as ABA-dependent TF regulator, recorded down regulation in *Arabidopsis thaliana* (62). Also reported that cationic peroxidase has the function to accumulate in the xylem vessels in rice plants.

Genome-wide analysis of chickpea genotypes against *Fusarium oxysporum* was done and transcriptome study conducted by illumining technology at conidial germination stage at variant points. The results revealed that; genes linked to fungal developments are transcribed at consecutive ways were discovered. It was also reported that genes related to secret effectors, cell wall degrading, metabolism, peptidases, and transporters-related enzymes were determined at the germination stage of conidial growth. Moreover, metabolism genes are upregulated at germination, while secondary metabolites and transporters genes were upregulated at a later stage (Sharma et al. 2016). The root structure and colonization (hypocotyl) and their expression profiling in infected genotypes and plant response factors were determined using two *Fusarium oxysporum*. The results revealed that less colonization in xylem vessels was recorded in weekly infected genotypes. After the analysis of virulent genes, the expression profiling results represent that two genes (SIX1, SIX6) include TF (FTF1) were upregulated in root crown and hypocotyl. Both strains performed differently, the virulent strain showed strong transcription in PR1 gene while other strains respond to ethyne factor ERF2 (Niño-Sánchez et al. 2015).

In general plant colonization by fungal vascular wilt pathogens after invasion colonization was done in cortical cells, latterly hyphae intercellularly move toward vascular parenchyma cells and occupied xylem vessels. Once reached to xylem, mycelium is restricted in the vessels; as a result necrosis occurs in host tissue for general colonization (Yadeta and Thomma 2013). Ma et al. (2010) also reported that *Fusarium oxysporum*-specific sequences present in replaceable chromosomal position are the basis of host specialization and polyphyletic origins of most formae specials.

4.8 Exclusion and Eradication of the Pathogen

The exclusion and eradication of the pathogen is the basic paradigm for crop improvement programs. For this purpose, integrated approaches have been used to exclude and eradicate crop diseases, pests, and weeds. Though disease control by the integrated management approach is no cure for plant disease control, it is considered as an ecology approach by which different disease control measures are adopting such as pathogen-free planting material, avoiding planting in high-risk soil, exclusion and eradication of *F. oxysporum* inoculum from rhizosphere, and using of biocontrol measures for healthy planting materials. It is transmitted through virulent seeds and plant residues (Jimenez Diaz et al. 2011; Nelson et al. 1981), infected materials than propagating into pathogen-free soils. For this purpose, strict legislation and inspection of the seeds material and planting area and optimize the use of FOC spp. in the non-virulent area (Jimenez Diaz et al. 2011). For quantification, evaluation, inspection, and legislation of the quarantine measurement, Jiménez-Fernández et al. (2011) established a qPCR protocol that permits to measure the DNA quantity in root and stems from infected asymptomatic chickpea. Seed dressing with Benlate could be used to remove seed borne inoculum (Haware et al. 1978).

Soil having problems of *Fusarium oxysporum* can be reclaimed by reducing or lessening the initial inoculum or reducing the disease potential (Passari et al. 2017; Jimenez Diaz et al. 2011), and this can be achieved by various methods, i.e., biological, physical, and chemical means. A most important method is soil solarization, and *Fusarium* wilt can be controlled in many crops in this way (Stapleton and de Vay 1986). By solarization, pathogen not only kills but also weakens and reduces the severity and increases the availability of other components in soil microbiota (Strange 2003). Moreover, soil pathogen can also be controlled by flooding (Strange 2003), by removing the plant residue from wilt affected crop, by killing the FOC chlamydospore, and by limiting the severity of the disease for the next crop (Jiménez Díaz et al. 2015). During biological control, use bio-agents to reduce the pathogen activity by making colonization in the rhizosphere while no toxic residue remains in the soil (Dubey et al. 2007). *Trichoderma* has been used against *Fusarium* wilt in greenhouse and field condition and gives tremendous result to control the disease (Kaur and Mukhopadhyay 1992).

Moreover, the application of *Pseudomonas* restricted the FOC in vitro and allowed significant growth in shoot length, dry weight, and yield (Nautiyal 1997). Application of nonpathogenic type strains such as *Bacillus* sp. and *Pseudomonas* recorded a significant reduction in the severity of *Fusarium oxysporum* f. sp. *ciceris* (Nautiyal 1997). Another practice could also reduce the severity of plant pathogen effect on the chickpea crop. An adequate amount of cultural practices takes the benefit of *Fusarium* management. A study reported that *Fusarium* can live about 6 years and 3 years of crop rotation but is not effective to reduce the effect of *Fusarium* incidence (Haware et al. 1996). Moreover, widespread disease development is due to the sowing date (Navas-Cortés et al. 1998); sowing chickpea from early spring to early winter could slow the *Fusarium* wilt development and ultimately enhance the yield (Landa et al. 2004). Along with the sowing date, the use of resistant cultivars also appears to be benefitted to control the wilt disease. Resistant varieties played an important role in an integrated disease management program (Landa et al. 2004; Jimenez Diaz et al. 2011; Jiménez Díaz et al. 2015). Resistant desi genotypes have been identified against FOC that reduced the disease incidence in wild and desi chickpea varieties (Jiménez Díaz et al. 2015). The availability of high genetic diversity in pathogenicity reduces the effectiveness and extensive use of present resistance (Bayraktar and Dolar 2012).

4.9 Conclusion

Fusarium oxysporum f. sp. *ciceri* (FOC) affects the chickpea crop causing wilt disease, more damaging and worldwide in occurrence. The main reason of *Fusarium* wilt is soilborne pathogen and showed symptoms, i.e., delaying crown, leaf anomalies, and rolled brown leaves. The number of strains is unknown of the soilborne pathogen and is difficult to control without solid information and identification of the pathogen. In general, symptoms of the disease occur at any stage of plant growth while more visible at the early stage of flowering and appears at the podding stage (late wilt). Late wilted plants exhibited falling of petioles, rachis, and leaflets as well as necrosis and discoloration of foliage. Early *Fusarium* wilt affects more than late wilting. However, late wilted plants produce lighter, rougher, and duller seeds as compared to normal. During the defense mechanisms, the plant uses many molecular signals or protein receptors to know the presence of microbes. Two modes of pathogen recognition are used by the host, i.e., effector-triggered immunity and pathogen-triggered immunity (PTI). The invariant epitope types are called microbe-associated molecular patterns and are composed of flagellin, chitin, and lipopolysaccharides that help spread the disease. Studies also reported that the sensing of bacteria produce siderophores and fungi serve as MAMPs and hydroxyproline and rapid alkalization factors, but their role was not clear yet in defense mechanism. The severity of the chickpea wilt is depending upon the pathogen, genotypes, pathogenic races, inoculum density, environmental condition, and cultivar sensitivity. The activity of the wilting disease was triggered by a combination of pathogen activities.

Breeding approaches involved genetic diversity the most important step for a breeding program, selection of desirable plants for trait resistance and disease resistance and evaluation of the plant for commercial production. In an intraspecific hybrid program, the single-cross method was used in desi and Kabuli chickpea genotypes with variant genetic history. Molecular markers are an important tool for identification, characterizing, screening, and diversity among the pathogens and diseases. Commonly, internal transcribed spacer (ITS) markers are used for classification and screening of the fungi. *While data regarding pathogen diversity is compulsory to comprehend pathology and development for control measures,* SSR markers are used in separate backcross generation to assist in the selection against the resistance of *Fusarium*. Many pathogenic FOC spp. cause alike symptoms in chickpea crop as with FOC. For this purpose, screening, identification, and insight are more important among the pathogen FOC spp. This approach provides a deep understanding of the epidemiology of the disease and triggered the development of elite resistant genotypes by adopting breeding, molecular, and plant omics technology. QTLs linked molecular markers would also facilitate to identify the desired traits is the basic requisite for the application of molecular markers in the breeding program and enhance the selection process. Combinations of SCAR with a codominant marker (CaETR) linked with QTLAR1 for *Ascochyta* tagging and help to identify the resistance alleles from a core collection of resistant cultivars. Moreover, NILs are used as a value able tool for mapping, refining the target region and selection of the desired gene for resistance to FOC 0. Efforts have been made to construct the genetic map using molecular markers for tagging traits and site-specific gene of interest in chickpea. However, the population derived from interspecific crosses was made due to microsatellite markers exploiting more genetic polymorphisms. Recent advances in molecular plant sciences boost the knowledge, and transcriptomic emerged as a powerful method to understand differential genic response over specific time-bound fashion. TF has the function to regulate the cell development, differentiation, and growth by tagging specific site with DNA or multiple sites and triggered the activation or repression of the TF through various mechanisms and interactions, i.e., DNA-protein, protein-protein, and alteration in chromatin structure. The infection includes root identification, colonization, penetration, adhesion, and penetration of the root cortex and hyphal proliferation within the xylem vessels are controlled by transcription factors (TFs). The functional characterization of the genes would also facilitate resistance mechanisms and their use in the breeding program against wilt resistance and crop tolerance mechanism along with target pathogen identification for the facilitation of the development of novel control management strategy.

Acknowledgments The authors would like to extend their sincere appreciation to the Deanship of Scientific Research at King Saud University for funding this research group NO (RG-1435-014).

References

Ahmad Z, Mumtaz AS, Ghafoor A, Ali A, Nisar M. Marker Assisted Selection (MAS) for chickpea *Fusarium oxysporum* wilt resistant genotypes using PCR based molecular markers. Mol Biol Rep. 2014;41:6755–62.

Ali L, Azam S, Rubio J, Kudapa H, Madrid E, Varshney RK, Castro P, Chen W, Gil J, Millan T. Detection of a new QTL/gene for growth habit in chickpea CaLG1 using wide and narrow crosses. Euphytica. 2015;204:473–85.

Allahverdipoor KH, Bahramnejad B, Amini J. Selection of molecular markers associated with resistance to *Fusarium* wilt disease in chickpea (*Cicer arietinum* L.) using multivariate statistical techniques. Aust J Crop Sci. 2011;5:1801–9.

Al-taae AK, Hadwan HA, SAE AJ. Physiological Races of *Fusarium oxysporum* f. sp. *ciceris* in Iraq. J Life Sci. 2013;7:1070–5.

Ashraf N, Ghai D, Barman P, Basu S, Gangisetty N, Mandal MK, Chakraborty N, Datta A, Chakraborty S. Comparative analyses of genotype dependent expressed sequence tags and stress-responsive transcriptome of chickpea wilt illustrate predicted and unexpected genes and novel regulators of plant immunity. BMC Genomics. 2009;10:415.

Bayraktar H, Dolar FS. Pathogenic variability of *Fusarium oxysporum* f. sp. *ciceris* isolates from chickpea in Turkey. Pak J Bot. 2012;44:821–3.

Beckman CH, Roberts EM. On the nature and genetic basis for resistance and tolerance of fungal wilt diseases. Adv Bot Res. 1995;21:35–77.

Beckman CH. The nature of wilt diseases of plants. St. Paul, MN: American Phytopathological Society; 1987.

Berrada AF, Shivakumar BG, Yaburaju NT. Chickpea in cropping systems. In: Yadav SSS, Redden R, Chen W, Sharma B, editors. Chickpea breeding and management. Wallingford: CABI Publishing; 2007. p. 193–212.

Bishopt GD, Cooper RM. An ultrastructural study of root invasion in three vascular wilt diseases. Physiol Plant Pathol. 1983;22:15–27.

Brown DW, Busman M, Proctor RH. *Fusarium verticillioides SGE1* is required for full virulence and regulates expression of protein effector and secondary metabolite biosynthetic genes. Mol Plant-Microbe Interact. 2014;27:809–23.
Castro P, Pistón F, Madrid E, Millán T, Gil J, Rubio J. Development of chickpea near-isogenic lines for *Fusarium* wilt. Theor Appl Genet. 2010;121:1519–26.
Castro P, Rubio J, Millán T, Gil J, Cobos MJ. *Fusarium* wilt in chickpea: general aspect and molecular breeding. In: Rios TF, Ortega ER, editors. *Fusarium*: epidemiology, environmental sources and prevention. New York, NY: Nova Science Publishers; 2012. p. 101–22.
Cha JY, Han S, Hong HJ, Cho H, Kim D, Kwon Y, et al. Microbial and biochemical basis of a Fusarium wilt-suppressive soil. ISME J. 2016;10:119–29.
Chand H, Khirbat SK. Chickpea wilt and its management-a review. Agric Rev. 2009;30:1–12.
Cobos MJ, Fernández MJ, Rubio J, Kharrat M, Moreno MT, Gil J, Millán T. A linkage map of chickpea (*Cicer arietinum* L.) based on populations from Kabuli × Desi crosses: location of genes for resistance to *Fusarium* wilt race 0. Theor Appl Genet. 2005;110:1347–53.
Cobos MJ, Rubio J, Fernández-Romero MD, Garza R, Moreno MT, Millán T, Gil J. Genetic analysis of seed size, yield and days to flowering in a chickpea recombinant inbred line population derived from a Kabuli × Desi cross. Ann Appl Biol. 2007;151:33–42.
Ding Z, Li M, Sun F, Xi P, Sun L, Zhang L, Jiang Z. Mitogen-activated protein kinases are associated with the regulation of physiological traits and virulence in *Fusarium oxysporum* f. sp. *cubense*. PLoS One. 2015;10:e0122634.
Dubey SC, Suresh M, Singh B. Evaluation of Trichoderma species against *Fusarium oxysporum* f. sp. *ciceris* for integrated management of chickpea wilt. Biol Control. 2007;40:118–27.
Garcia-Limones C, Hervás A, Navas-Cortés JA, Jiménez-Diaz RM, Tena M. Induction of an antioxidant enzyme system and other oxidative stress markers associated with compatible and incompatible interactions between chickpea (*Cicer arietinum* L.) and *Fusarium oxysporum* f.sp. *ciceris*. Physiol Mol Plant Pathol. 2002;61:325–37.
Gaur PM, Gowda CLL, Knights EJ, Warkentin T, Acikoz N, Yadav SS, Kumar J. Breeding achivements. In: Chickpea breeding. Wallingford: Centre for Agriculture and Bioscience International (CABI); 2007.
Gaur PM, Jukanti AK, Varshney RK. Impact of genomic technologies on chickpea breeding strategies. Agronomy. 2012;2:199–221.
Gaur PM, Slinkard AE. Genetic control and linkage relations of additional isozyme markers in chick pea. Theor Appl Genet. 1990;80:648–56.
Gowda SJM, Radika P, Kadoo NY, Mhase LB, Gupta VS. Molecular mapping of wilt resistance genes in chickpea. Mol Breed. 2009;24:177–83.
Gupta S, Bhar A, Chatterjee M, Das S. *Fusarium oxysporum* f. sp. *ciceri* race 1 induced redox state alterations are coupled to downstream defense signaling in root tissues of chickpea (*Cicer arietinum* L.). PLoS One. 2013;8:e73163.
Gupta S, Chakraborti D, Rangi RK, Basu D, Das S. A molecular insight into the early events of chickpea (*Cicer arietinum*) and *Fusarium oxysporum* f. sp. *ciceri* (Race 1) interaction through cDNA-AFLP analysis. Phytopathology. 2009;99:1245–57.
Gupta S, Chakraborti D, Sengupta A, Basu D, Das S. Primary metabolism of chickpea is the initial target of wound inducing early sensed *Fusarium oxysporum* f. sp. *ciceri* race 1. PLoS One. 2010;5:e9030.
Gurjar GS, Giri AP, Gupta VS. Gene expression profiling during wilting in chickpea caused by *Fusarium* oxysporum f.sp. *Ciceri*. Am J Plant Sci. 2012;3:190–201.
Halila I, Cobos MJ, Rubio J, Millán T, Kharrat M, Marrakchi M, Gil J. Tagging and mapping a second resistance gene for *Fusarium* wilt race 0 in chickpea. Eur J Plant Pathol. 2009;124:87–92.
Haware MP, Nene YL, Natarajan M. The survival of *Fusarium oxysporum* f. sp. *ciceri* in the soil in the absence of chickpea. Phytopathol Mediterr. 1996;35:9–12.
Haware MP, Nene YL, Rajeshwari R. Eradication of *Fusarium oxysporum* f. sp. *ciceri* transmitted in chickpea seed. Phytopathol. 1978;68:1364–7.
Haware MP, Nene YL. Influence of wilt and different growth stages on yield loss in chickpea. Trop Grain Legum Bull. 1980;19:38–40.

Haware MP. *Fusarium* wilt and other important diseases chickpea in the Mediterranean area. In: Saxena MC, Cubero JI, Wery J, editors. Present status and future prospects of chickpea crop production and improvement in the Mediterranean countries. Zaragoza: CIHEAM; 1990. p. 61–4.

Hiremath PJ, Farmer A, Cannon SB, Woodward J, Kudapa H, Tuteja R, et al. Large scale transcriptome analysis in chickpea (*Cicer arietinum* L.), an orphan legume crop of the semiarid tropics of Asia and Africa. Plant Biotechnol J. 2011;9:922–31.

Hiremath PJ, Kumar A, Penmetsa RV, Farmer A, Schlueter JA, Chamarthi SK, et al. Large scale development of cost effective SNP marker assays for diversity assessment and genetic mapping in chickpea and comparative mapping in legumes. Plant Biotechnol J. 2012;10:716–32.

Imtiaz M, Materne M, Hobson K, van Ginkel M, Malhotra RS. Molecular genetic diversity and linked resistance to ascochyta blight in Australian chickpea breeding materials and their wild relatives. Aust J Agric Res. 2008;59:554–60.

Iruela M, Rubio J, Barro F, Cubero JI, Millán T, Gil J. Detection of two quantitative trait loci for resistance to ascochyta blight in an intra-specific cross of chickpea (*Cicer arietinum* L.): development of SCAR markers associated with resistance. Theor Appl Genet. 2006;112:278–87.

Jendoubi W, Bouhadida M, Millan T, Kharrat M, Gil J, Rubio J, Madrid E. Identification of the target region including the Foc0 1 /foc0 1 gene and development of near isogenic lines for resistance to *Fusarium* Wilt race 0 in chickpea. Euphytica. 2016;210:119–33.

Jimenez Diaz RM, Jimenez Gasco MM, Landa BB, Castillo P, Navas-Cortes JA. *Fusarium* wilt of chikpea. In: Chen W, Sharma HC, Muehlbauer FJ, editors. Compendium of chikpea and lentil diseases and pests. St. Paul, MN: The American Phytopathological Society; 2011.

Jimenez-Diaz RM, Alcala-Jimenez AR, Hervas A, Trapero-Casas JL. Pathogenic variability and host resistance in the *Fusarium oxysporum* f. sp. *ciceris/Cicer arietinum.* Pathosystem. In: Proceedings of the 3rd European Seminar on *Fusarium* mycotoxins, taxonomy, pathogenicity and host resistance. Plant Breeding and Acclimatization Institute, Radzikov, Poland; 1993, pp. 87–94.

Jiménez Díaz RM, Castillo P, del Mar Jiménez Gasco M, Landa BB, Navas Cortés JA. *Fusarium* wilt of chickpeas: biology, ecology and management. Crop Prot. 2015;73:16–27.

Jiménez-Fernández D, Montes-Borrego M, Jiménez-Díaz RM, Navas-Cortés JA, Landa BB. In Plant and soil quantification of *Fusarium oxysporum* f. sp. *ciceris* and evaluation of *Fusarium* wilt resistance in chickpea with a newly developed quantitative polymerase chain reaction assay. Phytopathology. 2011;101:250–62.

Jimenez-Fernandez D, Navas-Cortes JA, Montes-Borrego M, Jimenez-Diaz RM, Landa BB. Molecular and pathogenic characterization of *Fusarium redolens*, a new causal agent of *Fusarium* yellows in chickpea. Plant Dis. 2011;95:860–70.

Jiménez-Gasco M, Pérez-Artés E, Jiménez-Diaz RM. Identification of Pathogenic Races 0, 1B/C, 5, and 6 of *Fusarium oxysporum* f. sp. *ciceris* with Random Amplified Polymorphic DNA (RAPD). Eur J Plant Pathol. 2001;107:237–48.

Kachuei R, Yadegari MH, Safaie N, Ghiasian A, Noorbakhsh F, Piranfar V, Rezaie S. PCR-RFLP patterns for the differentiation of the *Fusarium* species in virtue of ITS rDNA. Curr Med Mycol. 2015;1:4–11.

Kaiser WJ, Alcala-Jimenez AR, Hervas-Vargas A, Trapero-Casas JL, Jiménez-Díaz RM. Screening of wild Cicer species for resistance to races 0 and 5 of *Fusarium oxysporum* f. sp. *ciceris*. Plant Dis. 1994;78:962–7.

Kaur NP, Mukhopadhyay AN. Integrated control of chickpea wilt complex by trichoderma and chemical methods in India. Trop Pest Manag. 1992;38:37–41.

Kazan K, Muehlbauer FJ, Weeden NE, Ladizinsky G. Inheritance and linkage relationships of morphological and isozyme loci in chickpea (*Cicer arietinum* L.). Theor Appl Genet. 1993;86:417–26.

Kohli D, Joshi G, Deokar AA, Bhardwaj AR, Agarwal M, Agarwal SK, Srinivasan R, Jain PK. Identification and characterization of wilt and salt stress-responsive microRNAs in chickpea through high-throughput sequencing. PLoS One. 2014;9:e108851.

Kumar J, Choudhary AK, Solanki RK, Pratap A. Towards marker-assisted selection in pulses: a review. Plant Breed. 2011;130:297–313.

Kumar Y, Dholakia BB, Panigrahi P, Kadoo NY, Giri AP, Gupta VS. Metabolic profiling of chickpea-*Fusarium* interaction identifies differential modulation of disease resistance pathways. Phytochemistry. 2015;116:120–9.

Kumar Y, Zhang L, Panigrahi P, Dholakia BB, Dewangan V, Chavan SG, Kunjir SM, Wu X, Li N, Rajmohanan PR, Kadoo NY, Giri AP, Tang H, Gupta VS. *Fusarium oxysporum* mediates systems metabolic reprogramming of chickpea roots as revealed by a combination of proteomics and metabolomics. Plant Biotechnol J. 2016;14:1589–603.

Kusuya Y, Hagiwara D, Sakai K, Yaguchi T, Gonoi T, Takahashi H. Transcription factor Afmac1 controls copper import machinery in *Aspergillus fumigatus*. Curr Genet. 2018;63:777–89.

Landa BB, Navas-Cortés JA, Jiménez-Díaz RM. Integrated management of *Fusarium* wilt of chickpea with sowing date, host resistance, and biological control. Phytopathology. 2004;94:946–60.

Lev-Yadun S, Gopher A, Abbo S. The cradle of agriculture. Science. 2000;288:1602–3.

Lichtenzveig J, Bonfil DJ, Zhang HB, Shtienberg D, Abbo S. Mapping quantitative trait loci in chickpea associated with time to flowering and resistance to *Didymella rabiei* the causal agent of Ascochyta blight. Theor Appl Genet. 2006;113:1357–69.

Lucas J. Plant diseases. In: Plant pathology and plant pathogens. Malden, MA: Blackwell Publishing; 1998.

Ma LJ, van der Does HC, Borkovich KA, Coleman JJ, Daboussi MJ, Di Pietro A, et al. Comparative genomics reveals mobile pathogenicity chromosomes in *Fusarium*. Nature. 2010;464:367–73.

Madrid E, Chen W, Rajesh PN, Castro P, Millán T, Gil J. Allele-specific amplification for the detection of ascochyta blight resistance in chickpea. Euphytica. 2013;189:183–90.

Mendgen K, Hahn M, Deising H. Morphogenesis and mechanisms of penetration by plant pathogenic fungi. Annu Rev Phytopathol. 1996;34:367–86.

Millan T, Madrid E, Cubero JI, Amri M, Castro P, Rubio J. Chickpea. In: De Ron Antonio M, editor. Grain legumes. New York, NY: Springer; 2015.

Millan T, Winter P, Jüngling R, Gil J, Rubio J, Cho S, et al. A consensus genetic map of chickpea (*Cicer arietinum* L.) based on 10 mapping populations. Euphytica. 2010;175:175–89.

Nautiyal CS. Selection of chickpea-rhizosphere-competent *Pseudomonas fluorescens* NBRI1303 antagonistic to *Fusarium oxysporum* f. sp. *ciceri*, *Rhizoctonia bataticola* and *Pythium* sp. Curr Microbiol. 1997;35:52–8.

Navas-Cortés JA, Hau B, Jiménez-Díaz RM. Effect of Sowing Date, Host Cultivar, and Race of *Fusarium oxysporum* f. sp. *ciceri* on development of *Fusarium* wilt of chickpea. Phytopathology. 1998;88:1338–46.

Navas-Cortés JA, Hau B, Jiménez-Díaz RM. Yield loss in chickpeas in relation to development of *Fusarium* wilt epidemics. Phytopathology. 2000;90:1269–78.

Nelson PE, Tousson TA, Cook RJ. *Fusarium*: diseases, biology and taxonomy. University Park, PA: Pennsylvania State University Press; 1981.

Nelson PE. Life cycle and epidemiology of *Fusarium oxysporum*. In: Mace M, Bell AA, Beckman C, editors. Fungal wilt diseases of plants. London, UK: Academic Press; 2012. p. 51–78.

Niño-Sánchez J, Tello V, Casado-del Castillo V, Thon M, Benito EP, Díaz-Mínguez JM. Gene expression patterns and dynamics of the colonization of common bean (*Phaseolus vulgaris* L.) by highly virulent and weakly virulent strains of *Fusarium oxysporum*. Front Microbiol. 2015;6:1–34.

Rubio J, Hajj-Moussa E, Kharrat M, Moreno MT, Millan T, Gil J. Two genes and linked RAPD markers involved in resistance to *Fusarium oxysporum* f. sp. *ciceri* race in chickpea. Plant Breed. 2003;122:188–91.

Salimath PM, Toker C, Sandhu JS, Kumar J, Suma B, Yadav SS, Bahl PN. Conventional breeding methods. In: Yadav SS, Redden RJ, Chen W, Sharma B, editors. Chickpea breeding and management. Wallingford: Centre for Agriculture and Bioscience International (CABI); 2007.

Schippers B, Van Eck WH. Formation and survival of chlamydospores in *Fusarium*. In: *Fusarium*: diseases, biology and taxonomy. London, UK: The Pennsysylvania State University Press; 1981.

Shakir AS, Mirza JH. Location of seed-borne fungi in chickpea seed. Pak J Phytopathol. 1994;6:87–90.

Sharma KD, Chen W, Muehlbauer FJ. Genetics of chickpea resistance to five races of *Fusarium* wilt and a concise set of race differentials for *Fusarium oxysporum* f. sp. *ciceri*. Plant Dis. 2005;89:385–90.

Sharma KD, Muehlbauer FJ. *Fusarium* wilt of chickpea: physiological specialization, genetics of resistance and resistance gene tagging. Euphytica. 2007;157:1–14.

Sharma M, Sengupta A, Ghosh R, Agarwal G, Tarafdar A, Nagavardhini A, et al. Genome wide transcriptome profiling of *Fusarium oxysporum* f. sp. *ciceri* conidial germination reveals new insights into infection-related genes. Sci Rep. 2016;6:1–11.

Singh R, Sharma P, Varshney RK, Sharma SK, Singh NK. Chickpea improvement: role of wild species and genetic markers. Biotechnol Genet Eng Rev. 2008;25:267–314.

Singh S, Singh I, Kapoor K, Gaur PM, Chaturvedi SK, Singh NP, Sandhu JS. Chickpea. In: Broadening the genetic base of grain legumes. New Delhi, India: National Bureau of Plant Genetic Resources; 2014. p. 51–73.

Stapleton JJ, de Vay JE. Soil solarization: a non chemical approach for management of plant pathogens and pests. Crop Prot. 1986;5:190–8.

Stover RH. Banana root diseases caused by *Fusarium* oxysporum f. sp. *cubense*, *Pseudomonas solanacearum*, and *Radopholus similis*: A comparative study of life cycles in relation to control. In Root diseases and soil borne pathogens; 1970, pp. 197–200.

Strange RN. Introduction to plant pathology. London, UK: Wiley; 2003.

Thudi M, Bohra A, Nayak SN, Varghese N, Shah TM, Penmetsa RV, et al. Novel SSR markers from BAC end sequences, DArT arrays and a comprehensive genetic map with 1,291 marker loci for chickpea (*Cicer arietinum* L.). PLoS One. 2011;6:1–12.

Trapero Casas A, Jimenez Diaz RM. Fungal wilt and root rot diseases of chickpea in southern Spain. Phytopathology. 1985;75:1146–51.

Tullu A, Kaiser WJ, Kraft JM, Muehlbauer FJ. A second gene for resistance to race 4 of *Fusarium* wilt in chickpea and linkage with a RAPD marker. Euphytica. 1999;109:43–50.

Upadhyaya HD, Haware MP, Kumar J, Smithson JB. Resistance to wilt in chickpea. I. Inheritance of late-wilting in response to race 1. Euphytica. 1983;32:447–52.

Upasani ML, Gurjar GS, Kadoo NY, Gupta VS. Dynamics of colonization and expression of pathogenicity related genes in *Fusarium oxysporum* f.sp. *ciceri* during chickpea vascular wilt disease progression. PLoS One. 2016;11:1–21.

Upasani ML, Limaye BM, Gurjar GS, Kasibhatla SM, Joshi RR, Kadoo NY, Gupta VS. Chickpea-*Fusarium oxysporum* interaction transcriptome reveals differential modulation of plant defense strategies. Sci Rep. 2017;7:1–12. https://doi.org/10.1038/s41598-017-07114-x.

Varshney RK, Mohan SM, Gaur PM, Chamarthi SK, Singh VK, Srinivasan S, et al. Marker-assisted backcrossing to introgress resistance to *Fusarium* wilt race 1 and Ascochyta blight in C 214, an elite cultivar of chickpea. Plant Genom. 2014;7:1–11.

White TJ, Bruns T, Lee S, Taylor J. Amplification and direct sequencing of fungal ribosomal RNA genes for phylogenetics. In: Innis MA, Gelfand DH, Sninsky JJ, White TJ, editors. PCR protocols a guide to methods and applications. San Diego, CA: Academic Press; 1990. p. 315–22.

Yadeta KE, Thomma BPHJ. The xylem as battleground for plant hosts and vascular wilt pathogens. Front Plant Sci. 2013;4:1–13.

Yun Y, Zhou X, Yang S, Wen Y, You H, Zheng Y, et al. *Fusarium oxysporum* f. sp. *lycopersici* C2H2 transcription factor FolCzf1 is required for conidiation, fusaric acid production, and early host infection. Curr Genet. 2019;65(3):1–11.

Chapter 5
Integrated and Sustainable Management of Fungal Diseases of Chickpea: Current Status and Challenges

Babu Nagabhushan Motagi, M. S. Laxminarayan Rao, and Akshay Mathad

5.1 Introduction

Chickpea is an important commercial rabi pulse crop of the globe and India. India is a leading producer of chickpea ranked first both in an area with 99.27 lakh ha and production of 98.80 lakh tonnes of chickpea, followed by Pakistan, Iran, and Australia. The highest productivity of 3759 kg ha^{-1} is observed in China followed by Israel, the Republic of Moldova, and Bosnia and Herzegovina. However, Indian chickpea productivity is only 995 kg ha^{-1} (Anonymous 2016). The low productivity observed in India is mainly attributed to the increasing pests and diseases with poor management practices coupled with climate change. Chickpea crop is mainly affected by fungal diseases like *Fusarium* wilt (*Fusarium oxysporum.* f. sp. *ciceris*), ascochyta blight (*Ascochyta rabiei*), rust (*Uromyces ciceris-arientini*), dry root rot (*Rhizoctonia bataticola*), gray mold (*Botrytis cinerea*) and powdery mildew (*Leveillula taurica*), leaf spot (*Alternaria* sp.), phytophthora root rot (*Phytophthora medicaginis*), damping off (*Pythium debaryanum*), foot rot (*Sclerotium rolfsii*), and sclerotinia wilt (*Verticillium albo-atrum*).

Fusarium wilt is both soil and seed borne disease and is very hard to handle only by chemicals and also often breakdown of resistance owing to the presence of new virulent races, poses a true challenge for farmers and pathologists as a result of the scenario remains unchanged for last ten years, although attempts have been made in breeding and selecting several chickpea varieties with elevated disease-tolerant yield capacity. Epidemics of *Fusarium* Biodiversity can devastate plants and trigger

B. N. Motagi (✉) · A. Mathad
Department of Genetics and Plant Breeding, College of Agriculture
(University of Agricultural Sciences, Dharwad), Vijayapur, Karnataka, India
e-mail: motagibn@uasd.in

M. S. L. Rao
Department of Plant Pathology, College of Agriculture
(University of Agricultural Sciences, Dharwad), Vijayapur, Karnataka, India

B. P. Singh et al. (eds.), *Management of Fungal Pathogens in Pulses*,
Fungal Biology, https://doi.org/10.1007/978-3-030-35947-8_5

up to 100% losses in extremely infested areas and favorable circumstances. Resistant cultivars are the most efficient way of managing the disease and helping to stabilize the returns of chickpea. The development of *Fusarium*-resistant strains is focused primarily on standard choice in various breeding programs. This process takes time and relies on inoculum load and certain environmental influences on the growth of the disease. Using molecular techniques provides a good opportunity for enhancement of chickpea, in particular by defining molecular markers associated tightly to genes / QTLs that control *Fusarium* wilting (Warda et al. 2017). Biological control seems to be a better option and novel methods like 'Bio-priming' is being tested for the sustainable and eco-friendly management of diseases like *Fusarium* and *Sclerotium* wilt of both Chickpea (Vidhyasekaran and Muthamilan 1995). Ascochyta blight is the most serious disease causing up to 100% losses in Northern India, Pakistan, U.S.A. and Middle East (Smithson et al. 1985). Chickpea rust is also posing a serious threat and epidemics have been reported in several states like Karnataka, Andhra Pradesh, Maharashtra, etc. Further studies need to be carried out for a clear understanding of the biology of this pathogen, the role of alternate hosts like *Trigonella polycerata* survival of the pathogen in the, etc. Integrating bio-chemical monitoring seems to be an excellent way to combat many pathogenic agents with minimum intervention with the soil biological balance (Papavizas 1973).

Sequencing of reference genomes of CDC Frontier genotype in chickpea (Varshney et al. 2013a, b) and mapping of about 50 chickpea traits including blight, wilt, and gray mold diseases at ICRISAT helped in understanding the function of genes and pathways besides translating genomics research into product development in these important pulse crops. Superior chickpea line C 2014 with wilt and blight resistance is in multilocation trials for evaluation and release (ICRISAT 2017). Integrated and sustainable management of important fungal diseases of chickpea is discussed in the book chapter.

5.2 *Fusarium* Wilt

It is one of the most significant fungal diseases that can cause significant loss to chickpea crop worldwide. Butler first recorded it in India in 1918, but Padwick did not determine its etiology properly until 1940. The disease is now common in most of the Asian, African, Southern European, and American countries (Cunnington 2007). In India, it is widely distributed across Indo-Gangetic regions and elsewhere in Southern India.

5.2.1 *Symptoms*

The main symptom of *Fusarium* wilt in the field is drooping and the death of plants. The leaves turn yellow and drop off prematurely. In the wilted plants, necrosis of the collar region and discoloration are seen. The diseased plants can be easily removed from the soil, and most of the lateral roots are infected and become weak and remain in soil when plants are uprooted. The transverse section of the basal stem/roots revealed masses of hyphae under the microscope in the vascular bundles and discoloration of vascular cells.

The disease symptoms can be seen at any stage of the plant, and affected plants are in patches or spread across the whole field (Trapero-Casas and Jiménez-Díaz 1985). Sensitive cultivars may have signs of premature wilting, with flaccidity of individual plants and a dull green coloration following complete plant desiccation within 25 days after the sowing period. Late wilting signs, however, are generally most visible at flowering, and even appear until podding, when the petioles and leaflets drop, accompanied by yellowing and necrosis of foliage. In the upper part of the plant, drooping is seen first but occurs over the whole plant within a couple of days. Symptoms may only affect a few plant stalks that trigger partial wilting. The xylem of roots and stems develops dark-brown coloration and seen when made vertical/cross sections (Fig. 5.1). *Fusarium* decreases the production of chickpea by reducing both the yield and weight

Fig. 5.1 Field view of disease symptoms of major fungal diseases of chickpea and name of pathogens are marked in figure

(Nene and Haware 1980). Yield loss due to *Fusarium* wilt in India and Spain is 10–15% (Singh and Dahiya 1973) and 40% in Tunisia (Bouslama 1980) have been reported. Early wilting had greater yield reduction (77–94%) than yield reduction (24–65%) due to late wilting (Nene and Haware 1980).

5.2.2 Causal Organism

Fusarium oxysporum f. sp. *orthoceras* (Appel & Wollenweber) Bilay (class, Deuteromycetes; order, Moniliales; family, Tuberculariceae).

The fungus produces both inter- and intracellular hyaline mycelia in the infected tissue most abundantly in the vascular bundles. The fungus produces both macro- and microconidia in the host tissues as well as in cultures. Microconidia are small, thin-walled, hyaline, elliptical 1–2 celled, measuring 4–6 × 2–4 μm. Macroconidia are long, curved (fusiform or sickle-shaped) pointed at both ends, septate, and measure 25–40 × 3–4 μm. Chlamydospores, the surviving structures, are also formed in the host as well as in old cultures, which develop from any cell of the hypha. The cells round off and become thickly walled to form chlamydospores; they are spherical or oval single or, in chains, terminal or intercalary.

Although monophyletic, *F. oxysporum* f. sp. *ciceris* shows considerable pathogenic variation. Different pathogen syndromes with brown vascular discoloration were noticed depending upon the unique yellowing or wilting syndromes which make chickpea genotypes susceptible. Pathotypes, which are genetically diverse, are being placed in two separate groups depending on fingerprint assays RAPD, SCAR, and DNA (Jimenez-Gasco et al. 2001). Haware and Nene (1982b) reported that there is lot of variation in symptom types because of the presence of the eight races of pathogen (races 0, 1A, 1B/C, 2, 3, 4, 5, and 6), which were identified by reactions on a set of differential cultivars of chickpea. Races 0 and 1B/C induce the yellowing syndrome (yellowing pathotype), whereas races 1A, 2, 3, 4, 5, and 6 induce the wilting syndrome (wilting pathotype), and all the races have distinct geographic distributions. Haware and Nene (1982a) reported only four races (Races1A, 2, 3, and 4) in India, whereas races 0, 1B/C, 5, and 6 are found mainly in the Mediterranean region and the USA (Jimenez-Gasco et al. 2001). Three new races were reported from India based on old differentials (Honnareddy and Dubey 2006). The isolates from each state of India were highly variable, and based on the reactions on international differentials, more than one race were found to be prevalent in every state (Dubey and Singh 2008). Dubey et al. (2012), based on new differential set of chickpea cultivars, reported that all eight races were found in India.

5.2.3 Disease Cycle

The wilt-causing fungus survives saprophytically being facultative saprophyte and on dead organic matter in soil when the crop is harvested the diseased roots are left over in the soil. It also produces chlamydospores that survive in soil and becomes active in the next cropping season. The perfect stage is unknown.

5.2.4 Integrated Management

Fusarium wilt being both soil- and seed-borne is difficult to manage by chemical alone which may not be practically feasible. Accordingly, this calls for an integrated approach, involving chemical, biological, and genetic approaches. Several attempts have been made by several workers to manage this disease biologically.

Chickpea *Fusarium* wilt is mainly driven by the pathogen inoculum as it is a monocyclic disease. Therefore, its management should aim at excluding the pathogens and decreasing the original inoculum quantity by using measures like (i) pathogen-free seeds; (ii) avoiding sowing in disease-affected soils; (iii) elimination or reducing of soil inoculum; (iv) resistant varieties; (v) seed treatment with biocontrol agents or fungicides; and (vi) avoiding cropping patterns which favor infection by the pathogen (Jiménez-Díaz et al. 2015).

Bio-priming of seeds with *P. fluorescens* effectively controlled chickpea wilt disease in addition to increased yield. The seed treatment of *P. fluorescens* followed by its application in the root zone has not only increased the efficacy of *P. fluorescens* formulations but also enhanced the chickpea yields. *Pseudomonas fluorescens* does not have any adverse effect on the beneficial N-fixing bacteria, viz., *Rhizobium* and *Azospirillum*, and *P. fluorescens* were not inhibited by the thiram and carbendazim seed treatment fungicides (Vidhyasekaran and Muthamilan 1995).

The practical and cost-efficient individual measures for wilt management include developing and using high-yield cultivars which are resistant to the common pathogenic races(s) of *Fusarium* wilt in a specified region. *Fusarium* wilt management could be helped by the use of plants that do not have any pathogens (Pande et al. 2007), sanitary procedures and soil inoculum reductions, selection of sites, and attention to reducing the disease capacity and the protection of plants with fungicides. For the characterization and tracking of *Fusarium*, molecular protocols are accessible. In the course of the integrated management strategy, the improved management of these disease control interventions can be further achieved through mixing slow-wilting cultivars (Jiménez-Díaz et al. 2015).

Effective test fungicides, bioagent, and organic amendments were evaluated for integrated management of *Fusarium* wilt. The seed treatment with the combination of carbendazim, thiram, *Trichoderma viride*, and *P. fluorescens* followed by soil application with neem seed cake powder was found to be an effective treatment which resulted in significantly higher seed germination, lower incidence of wilt, and high seed yield compared to control treatment (Thaware et al. 2016).

There have been significant advances in identifying the desi and kabuli chickpea germplasm types and in developing productive high-performance 'Kabuli' cultivars with full resistance to more strains of the pathogen. There have also been substantial advances in the breakdown of racial resistance genes. This would allow further advancement in pyramiding of various strain-specific resistances in chickpea, which would increase the efficiency in multilocations and possibly merge this with resistance to other major diseases, viz., root-knot and cyst nematodes and blight, and tolerance to drought. But resistance hasn't been broken up to date by the use of racially specified resistant cultivars. Pre-planting of the existing pathogen with molecular protocols would assist to prevent the affected soils. In chickpea germplasm, slow-wilting resistance is also recognized. Increased effectiveness of the integrated wilt management in chickpea would be combined with other pre-planting disease control practices, viz., pathogen-free seeds, avoiding sowing in disease-affected soils, elimination or reducing of soil inoculum, resistant varieties, and seed treatment with biocontrol agents or fungicides, which would control the *Fusarium* wilt in chickpea.

Marker-assisted introgression was performed with foreground selection with SSR markers TA 37 and TA110 in Pusa 256 (elite desi cultivar) and with background selection with 45 SSRs accommodating 8 multiplexes to get the higher recovery of recurrent parent genome. Finally, there have been acquired 17 BC3F4 and 11 BC3F3 lines that have resulted in the detection of 5 high-resistance Pusa 256 strains with Foc 2 genes. This will assist the development of chickpea horizontally and vertically in India (Aditya Pratap et al. 2017).

5.3 Ascochyta Blight

It is the most important disease reported from 25 countries around the world (Singh et al. 1984) that includes Europe, North Africa (bordering Mediterranean Sea, Iran, Iraq, Pakistan, Portugal, Romania, Spain, the USA, USSR (formerly), Mexico, Tanzania, Bangladesh, and India, while it is not reported in chickpea areas of Nepal, Myanmar (Burma), Argentina, Bolivia, Peru, Chile, Libya, Columbia, Malawi, Zambia, Sudan, Uganda, and Yugoslavia. In India chickpea blight is common in Punjab, Haryana, Himachal Pradesh, Northwest Uttar Pradesh, and Bihar, Madhya Pradesh, but not from Andhra Pradesh. Recently its incidence has been observed from Karnataka state also. During the 1930s total loss due ascochyta blight in Spain was reported, and losses up to 25–50% were reported during 1922–1933 from undivided Punjab (before partition of Pakistan). In Rajasthan, 5–75% losses have been observed in 1982, under favorable environment disease severity increases resulting in losses up to 100%.

5.3.1 Symptoms

It occurs in all parts of the plant above ground. On the leaves and pods, circular spots develop and elongated spots on the petioles and stems. These leaf spots can have brown dots with a brown-red margin. On coalescence, the places turn whole to leaf gray with a scorched look. The lesions on green pods are curved and dark in the edges and are placed in a concentrated circle with pycnidia. In the clusters of seeds, lesions can also appear. In stems and petioles, the red with black dots are elongated that may cover the impacted area. The sections above these lesions drop out and die when such places girdle the stem entirely (Fig. 5.1). The whole plant dries when the main stem is located at the bottom (neck area). As the disease progresses, patches of drooping and wilting crops can subsequently spread to whole areas. The distribution may be limited in dry weather, but it extends quickly in moist conditions.

5.3.2 *Causal Organism*

Ascochyta rabiei (Pass.) Labrousse. Also referred to as *Phyllosticta rabiei* (Pass.) or *Phoma rabiei* (Pass.), the pathogen belongs to subdivision, Deuteromycotina; class, Coelomycetes; order, Sphaeriales; and family, Sphaeropsidaceae in which globose, dark pycnidia with hard textured walls are formed.

The pathogen produces hyaline to brownish septate mycelium. The pycnidia are produced on leaves, stem petioles, and pods including seeds which are erumpent, globose dark brown 140–200 μm in diameter with a prominent ostiole. The perfect stage (observed in Bulgaria by Kovachevski in 1936) described as *Mycosphaerella rabiei* Kov. (later renamed as *Didymella rabiei* (Kov.)) belongs to the family, Dothideaceae; order, Dothideales; and class Loculoascomycetes of Ascomycotina. The pseudothecia (perithecia in locules) contain eight small ascospores, immersed in host tissues (dead parts or in crop debris) dark brown or black globose and measure 120–250 × 75–152 μm. They contain cylindrical-clavate asci slightly curved pedicellate which measures 48–70 × 9–13.7 μm. The ascospores are one septate and one cell is bigger than the other prominently formed at the septum and measure 12.5–19.0 × 6.7–7.6 μm; however, in Indian conditions, these perfect stages are not observed as hot summer conditions prevail after the cropping season.

5.3.2.1 Races

Based on the reactions of the cultivars, the population of *A. rabiei* were grouped into seven races, and differential cultivars for each race were identified. The isolates were also analyzed for their genetic diversity using ITS, URP, and SSR markers (Baite and Dubey 2015). The presence of races could not be found by Luthra and others in 1939 and by Arif and Jabbar (1965). An anonymous study from India (1963) indicated that genotype C-12/34 broken its resistance due to a new strain. In controlled environments, scientists examined variations in fungal isolates. Based on the symptoms, the pycnidial formation and the pathogenic behavior of the eleven isolates were found and several races exists in Panjab, India. Further, findings from the Chickpea International Ascochyta Blight Nursery were also indicated the presence of races. Intensive race studies are needed to identify stable host resistance (Nene, 1981).

5.3.3 *Disease Cycle*

Blight pathogen survives as pycnidia in seeds and plant debris that is a major source under Indian conditions. However, pycnidia survive for more than 2 years in crop debris depending on temperatures (10–35 °C) and RH 65–100%. The fungus survives on the seed coat, cotyledons, and embryo for >5 months.

The pathogen spreads from these sources (infected debris and seeds) by rain droplets in windy weather, by insects and contact between leaves, and by movement of animals through the field. The 22–26 °C temperatures and high rainfall conditions are conducive for disease development at all crop growth (seedling to pod formation) stages. The pathogen has been noticed on berseem also with cross inoculums from these counterpart hosts, besides common bean (*Phaseolus vulgaris*).

5.3.4 Integrated Management

Genetic resistance to ascochyta blight: The resistance to G-52 isolate of ascochyta blight in chickpea was under the control of single dominant gene pair in the I-13 resistant variety (Satya Vir et al. 1975).

There have been efforts to identify sources of resistance, resistance breeding, and genetic variability between the blight pathogen races. Importance of the genotype x environment interaction in elucidating aggressiveness of isolates from different places and identifying pathotypes and stable sources of resistance has been recognized. The current blast resistance breeding programs rely on crossing durable and adaptive cultivars, stable performance of breeding lines through multilocation testing, and the marker-assisted selection (Sharma and Ghosh 2016).

Molecular diversity analyses of Indian isolates of Ascochyta rabiei: About 11 AFLPs and 20 SSR markers were evaluated in 64 isolates obtained from various agroclimate areas in North Western Plains Zone (NWPZ) India for the study. Some 9 polymorphic AFLP primer pairs produced 317 fragments with a median PIC value of 0.28, 130 of them are polymorphic. Of the SSR markers, 12 were polymorphic and had an average PIC value of 0.35 with a total of 29 alleles. This is the first AFLP and SSR diversity assessments in *A. rabiei* in the best of our understanding. The dendrograms were created respectively and placed the series of AB isolates in geographical areas based on AFLP and SSR information and the merged variable dataset. The population structure assessment model disclosed that 4 separate populations of different concentrations of ancient admixtures were explored between 64 isolates. Interestingly, several SSR markers and AFLP primer combinations showed the locus/allele specific to AB isolates from certain regions, viz., Gurdaspur, Hisar, Sundarnagar, and Sriganganagar. Genetic variability found in Indian NWPZ AB isolates indicates that modifications in *A. rabiei* population should be monitored continuously to prevent the collapse of resistance in chickpea cultivars.

Management

- Good Agronomic Practices (GAP) such as deep plowing, deep sowing, removal and destruction of crop debris, and crop rotation need to be followed.
- Intercropping with cereals reduces the disease spread (chickpea-barley).
- Application of 40–60 kg potash +20 kg nitrogen +40 kg phosphorus was reported to reduce the disease severity and increase grain yield (Tripathi et al 1987).
- Seed treatment with copper sulfate, thiram, or Calixin M (this last named fungicide completely eradicates the seed inoculum). Tripathi et al. (1987) have

reported successful control, by seed treatment with carbendazim + thiram (1:3 ratio) @2.5 g/kg seeds followed by three times spraying of carbendazim @0.5 kg/ha at 10-day interval.

- In mild infections, spraying of zineb, ferbam, maneb, or captan and Daconil, Rovral, Calixin M, tebuconazole, difenoconazole, chlorothalonil (Bravo), or azoxystrobin (Amistar 250 SC) can be taken @0.1% to 0.2%. Four to six sprays may be required depending upon disease severity and stage of the crop.
- Use of resistant varieties (ICRISAT and other centers in the country) such as F-8, C 325, C 727, I 13, EC 26414, 26,435, and 26,446. The Kabuli types ILC 3664, 3870 and 4421, and C 215 have been reported to be resistant to blight. Generally, Kabuli types are more resistant than desi chickpea or gram. It has observed that the resistant genotypes are hairier than susceptible plants that produce more maleic acid than healthy plants. Erect growth, less lateral spread, high hairiness, high peroxidase activity, lesser maleic acid content, higher L-cystine, and phenolic contents are the attributes of resistant varieties.

5.4 Rust

The rust of gram is reported from >15 countries. The disease is widespread in several parts of India including Maharashtra, Tamil Nadu, Bihar, West Bengal, Uttar Pradesh, and Punjab and recently in many places of Karnataka.

5.4.1 Symptoms

The rust appears around 4-month-old crop (January–February) on small leaves and light-dark brown pustules which tend to coalesce to form bigger pustules which may develop on either side of the leaf preferably on the lower surface and covers the entire leaf area later. Often the pustules appear on the stem, petioles, pods, and floral parts. In advanced stages, dark telial stages appear in rust pustules (Fig. 5.1).

5.4.2 Causal Organism

Uromyces ciceris-arietini (Gregnon) Jacs. The pathogen was first detected and described in France in 1863. The pycnidial and aecial stages of rust pathogen are unknown. The uredia are hypophyllous, scattered minute round powdery when mature light brown. The urediospores are globose, loosely echinulate, 20–28 μm in diameter, and yellowish brown in color. The telia appear late in the season (March–April) and resemble uredia except for dark brown color. The teliospores are round or oval or warty or angular with a roundish unthickened apex. The wall is brown and warty and measures 18–30 × 8–24 μm with short hyaline pedicel.

5.4.3 Disease Cycle

Rust fungus will survive by repeating its uredial stage, while the role of telia is unknown. The pathogen is known to infect the legume weed *Trigonella polycerata* and *Lathyrus* spp. and collateral hosts of *Uromyces ciceris-arietini* on high hills in summer and provide inoculums to the main host. The disease is favored by temperatures of 11–30 °C; the uredospore germination takes place in leaf exudates of susceptible varieties as resistant varieties are low in leaf exudates. Leaves of resistant varieties Nandriyal 49 contain more of maleic acid and sucrose than susceptible genotype Agra local (Bahadur and Sinha 1970).

5.4.4 Integrated Management

Image processing of rust disease: Automatic plant disease detection is an important aspect, which can demonstrate advantages in the surveillance of wide crop areas and therefore automatically identify disease symptoms when they appear on plant leaves. Costs and inaccuracies may be the problem with sheer naked eye monitoring for the detection and classification of diseases. The model suggested offers a software alternative for the traditional techniques integrated into the identification of programmed crop diseases with the use of the picture handling method. This system is beneficial for farmers to control the disease spread. It also offers precise outcomes with naked eyes. The method commences with chickpea leaving from the field being captured. Captured pictures are filtered, and then the green pixels are disguised and deleted with a certain limit value. The complete area on the disease-affected leaf and the good region is calculated based on the result. Texture characteristics are finally obtained (Shivanand et al. 2014).

Rust resistance in chickpea germplasm collection: A collection comprising 140 chickpea lines and 109 related wild (*Cicer* spp.) species has been screened for chickpea rust resistance. Different levels of partial resistance have been identified based on reduced disease seriousness and disease progression area, curve, and host cell necrosis macroscopically visible. In wild *Cicer* species, higher rates of resistance but not linked with hypersensitivity were found macroscopically and microscopically, and resistant components were researched in chosen *C. arietinum* accessions. During the long latent period a reduced infection was expressed that are associated with a greater percentage of early colonies aborted, a decrease in the amount of haustorial colony and mother cell and a reduction in the size of the colony (Sillero et al. 2012).

Management

1. Early sowing is known to provide disease escape mechanism.
2. Many antagonistic fungi suppress spore germination of *Uromyces ciceris-arietini*.

3. Growing rust resistant chickpea line viz., NRC 34, NEC 249, JM 583, and 2649, HPC 63, HPC 136 and HPC 147 is recommended in rust epidemic area.

5.5 Dry Root Rot of Chickpea

5.5.1 Symptoms

This disease usually occurs as scattered dead plants around flowering and podding time. Petioles and leaflets are drooped at the bottom of the plant. Uppermost leaves are chlorotic when the remaining are dry on the plant. The taproot is pale and has indications of drying, and most of its lateral and finer branches are empty.

5.5.2 Causal Organism

In the altered climate situation, chickpea dry root rot induced by *Rhizoctonia bataticola* (*Macrophomina phaseolina*) is gaining significance when increasing crops are exposed to elevated temperature and water stress. Many soil and climate variables are accountable for disease growth as these are primarily soilborne pathogens. So far, there has been no systemic ecological, biological, and epidemiologic study linked to dry root rot in chickpea. Investigations are required to enhance the characterization and identification of variation within its pathological and epidemiological niches. A limited accessible manuscript on HPR of dry root rot indicates that the disease has no resistant sources (Sharma et al. 2016).

The DRR was initially reported by Mitra (1931) in India subsequently, in Iran (Kaiser et al. 1968), the USA (Westerlund et al. 1974), and several Asian and African countries (Nene et al. 1996). The disease was formally recognized in chickpea as "rhizoctonia wilt," but was subsequently called as "dry root rot." In the recent years, changes in weather conditions, especially owing to a long drought, it has become an extremely serious risk to chickpea production. Chickpea is predisposed to DRR utilizing elevated temperature and depletion of soil water during plant development, especially post-harvest stages (Sharma and Pande 2013). The wide and enhanced prevalence of DRR in Central and Western India was stated in recent 2010–2013 studies (Ghosh et al. 2013). Regardless of soil, cultivars, and cropping systems, diseases were detected, and their prevalence ranged from 5 to 50% in poorly affected soils.

Dry root rot is an important biotic limitation for chickpea production. A total of 94 isolates from various agroclimate areas of India were analyzed with AFLP. Distinct morphological characteristics were evaluated to identify the variety of *Rhizoctonia bataticola* species in India. *Rhizoctonia bataticola* species were varied in terms of distinct moral and cultural parameters from various agroecological areas such as colony color, development pattern, development frequency, mycelial characteristics, sclerotial intensity, sclerotial initiation time, and sclerotial morphology. A total of 121 fragments were obtained from five AFLP primer combinations. All fragments were found to be polymorphic with an average value of 0.213 for polymorphic data content. Based on AFLP assessment, the dendrogram found that the highest amount of isolates of *Rhizoctonia bataticola* was varied and did not rely on geographical origin. Morphological and molecular information linked and endorsed the diversity and independence of the *Rhizoctonia bataticola* found in India (Sharma et al. 2012).

Dry root rot external expression: phenotypical modification: DRR signs are most often seen in the afterblown phase of chickpeas, which includes drooping and chlorosis of leaflet which is restricted to top plant leaves. The plant leaf and stalks are generally straw-colored and the reduced branches and stalks are gray in some cases. The root of the tap has red symbols that become black and absent in most of the lateral and softer components. The radicals that died are quite fragile and bark tipped. The roots revealed and the inner part of the bark, or when divided up vertically on the collar region, are observed with dark sclerotic minute bodies (Sharma et al. 2016).

5.5.3 Disease Cycle and Histopathology

DRR is usually caused by the presence of hyphae and sclerotia in the soilborne inoculum. The pathogen creates epidermal cell death and penetrates the roots. Mechanical plugs of the xylem cells by micro-sclerotia, enzyme action, toxin production, and mechanical stress lead to disease development and direct secretion of macerating enzymes (Sharma et al. 2004). The pathogen may also cause disease during the formation of cotyledons, through tiny rootlets/injuries on the root surface. The fungus develops within the cell as well as between the cells of cortical tissue. It mainly grows intercellular, forming thick and dark-colored cells which lead to large necrotic lesions that are depressed. Invaded cortical cells cause the roots to decay or to rot severely (Singh and Mehrotra 1982). The vascular system and the sclerotic bodies of the pathogen are colonized by hyphae. The level of root necrosis rises gradually over time without obvious signs in the above ground until blooming and podding.

5.5.4 Integrated Management

Host plant resistance: There have been so far researches on DRR resistance in chickpea, as neither demonstrated significant resistance to DRR. Comprehensive list of scientists worked on DRR resistance breeding and their findings on sources of DRR resistance/tolerance (Table 5.1).

Table 5.1 Resistance sources for dry root rot disease in chickpea

Chickpea lines	DRR disease reaction	Reference
GCP-101, GBM-2, GBM-6, and ICCV-10	Tolerant	Jayalakshmi et al. (2008)
ICCV-97112	Resistant	Iftikhar and Ilyas (2000)
ICCV-05530, ICCV-08305, ICCV-05529, ICCV-05532, ICCV-07117, and ICCV-07112.	Moderately resistant	Sharma et al. (2016)

Table 5.2 Important cultural practices to avoid the DRR incidence in chickpea

Sl. no	Cultural method	Reference
1	Manipulation in the date of sowing, i.e., timely or early sowing followed by scheduled irrigation can avoid the elevated temperatures thereby reducing the DRR	Singh et al. (1990)
2	Crop rotation with non-host crop plants	Singh et al. (1990)
3	No-tillage	–
4	Deep plowing and removal of infected debris for the reduced sclerotial multiplication	–

Table 5.3 Important biological control measures to avoid the DRR incidence in chickpea

Sl. no.	Biological control	Reference
1	Seed treatment with *Trichoderma viride*	Sharma and Gupta (2004)
2	Application of antagonistic *Trichoderma virens* and organic amendments like FYM	Thilagavathi et al. (2007)
3	A combination of biocontrol agents, viz., *T. viride*, *Pseudomonas fluorescens*, and *Bacillus subtilis*	Thilagavathi et al. (2007)

Inheritance of DRR resistance: DRR resistance inheritance study reveals that it is controlled by dominant monogenic genes (Rao and Haware 1987), in which two resistant (H-208 and K-850) and two sensitive relatives (C-104 and P-165) have been used. Even resistant parents had signs of the disease if the crops were cultivated in infected soil for a longer period. More refinement of screening techniques is needed as well as further confirmation of resistance sources in regulated environments and field. This brings the breeding of chickpea resistance to DRR in a scenario of uncertainty, especially now when the climate is unsafe. Besides, no study has been recognized so far regarding any molecular markers associated with the DRR gene.

Cultural control: The incidence rates of the disease can be decreased by cultural methods as listed below that can lead to a decreased occurrence of DRR (Table 5.2).

Biological control: Some of the important biological control measures are given in Table 5.3.

Chemical control: Seed treatment with fungicide is effective in reducing the losses due to *Rhizoctonia bataticola*. Some of the chemical control measures are listed in Table 5.4.

Table 5.4 Important chemical control measures to avoid the DRR incidence in chickpea

Sl. no.	Seed treatment chemicals	Reference
1	Carbendazim, thiophanate-methyl, and Vitavax	Sharma and Gupta (2004)
2	Carbendazim or in combination with thiram (soil drench and seed treatment + drenching after sowing)	Sharma and Gupta (2004
3	Bavistin and thiram	Ghosh et al. (2013)

5.6 Botrytis Gray Mold

5.6.1 Symptoms

At any stage of development, plants can be targeted by the pathogen gray mold, which is most probably found at the bottom of the stalk of the collar region as soft rot. In the beginning, the tissues in the injured condition are coated with a fuzzy gray mold, and as the disease develops, plants will be desiccated and die. On the surface of the affected tissue, small black sclerotia can occur when the plant dies. In older crops, only a few parts of the plant are occasionally damaged, and the remaining appears to be quite regular. The disease with seedlings can trigger damping and significant thinning.

5.6.2 Causal Organism

A fungus known as *Botrytis cinerea* causes this disease. The disease can grow quickly, distributed extensively, and trigger a complete loss of yield under favored circumstances. Genotypes of chickpea with strong seedling development, early flowering, and early canopy closure are amenable for disease development compared to other varieties. Total crop failure is reported during the use of heavily infected seeds and when seed treatment is not followed in some cases. Crop losses in moist periods are highest, especially when plants are developing very thick canopies.

5.6.3 Disease Cycle

As soilborne sclerotia and saprophyte grow on decaying crop waste, the fungus survives on infected plants. The disease often occurs through the sowing of infected seeds in fresh fields. On infected crops, masses of spores are generated. The fungal spores can be transmitted through air currents from one crop to another crop and distributed quickly. The hot, damp circumstances under the plant canopy offer perfect circumstances for infection and disease propagation once the plant has established. Botrytis Grey Mould management in chickpea is indicated in Table 5.5.

Table 5.5 Botrytis Grey Mould management in chickpea

Sl. no.	Methods	Practices
1	Cultural method	By use of disease-free seeds Low seed rates Wider row spacing
2	Biocontrol method	Soil or seed or foliar treatment of *Trichoderma harzianum*

5.7 Other Minor Fungal Diseases of Chickpea

*5.7.1 Powdery Mildew (*Leveillula taurica*)*

Symptoms

- Oidiopsis type of powdery mildew in which the mycelium is endophytic.
- The affected leaf shows powdery patches on the lower surface corresponding with yellowing on the upper surface.
- Older leaves show symptoms first.
- There will be premature defoliation of affected leaves.
- Airborne

Management

Spray carbendazim 1 g/lit or carbendazim + mancozeb (1 g/it) or wettable sulfur 2.5 g/lit.

*5.7.2 Blight (*Alternaria alternata*)*

Symptoms

- The disease occurs at flowering stage.
- Leaves are infected most.
- Shedding of infected lower leaves.
- Small, circular, water-soaked, and purple lesions are seen on leaflets
- Infected pods become blackish and seeds shriveled.

Management

- Space planting
- Reduced vegetative growth
- Intercrop with linseed
- Limited irrigation
- Compact varieties
- Mancozeb @ 2.5 g/lit or carbendazim @ 1 g/lit

5.8 Host Plant Resistance and Molecular Markers for Major Fungal Diseases of Chickpea

The chickpea has a limited genetic base and often does not have sources of resistance to several stresses including major fungal diseases in the cultivated germplasm. Thus, it is critical for the development of different cultivars to diversify and broaden the genetic base using wild relatives. In the past, some attempts have been done to monitor germplasm samples for valuable DNA to resist ascochyta blight, *Fusarium* wilt, botrytis gray mold, and other diseases under field and controlled circumstances. Through such attempts, precious resistance sources have been identified to these major fungal diseases in chickpea (Table 5.6). Efforts to develop genomic resources resulted in the identification of molecular markers for agronomic and biotic stresses, enabling the use of genomics-assisted breeding in chickpea crop (Varshney et al. 2013a). In the recent past, marker-assisted selection tool using SSR and SNP resources and density genetic map of chickpea have significantly augmented the chickpea breeding programs effectively and efficiently (Varshney et al. 2010; Kumar et al. 2011). Furthermore, genome sequencing of 90 chickpea has accelerated the development of disease resistance lines from molecular breeding efforts (Varshney et al. 2013b). However, this has some limitations, viz., not all the genes or QTLs for major diseases are fine-mapped and new sources of resistance need to be genotyped (Zhu et al. 2008).

Table 5.6 Resistance/tolerance sources to major fungal diseases of chickpea

Major disease	Resistance sources	References
Ascochyta blight	ILC 72, ILC 191, ILC 196, ILC 201, ILC 202, ILC 2506, ILC 2956, ILC 3274, ILC 3279, ILC 3346, ILC 3856, ILC 3956, ILC 3996, ILC 4421, ICC 3634, ICC 4200, ICC 4248, ICC 4368, ICC 5124, ICC 6981, ILWC 7–1, ILWC 33/S-4, 03039, 03041, 03053, 03115, 03131, 03133, 03143, 03159, 93A-086, 93A-111, 93A-3354	Malhotra et al. (2003) Ilyas et al. (2007) Kumar et al. (2011)
Fusarium wilt	JG 16, JG 62, ILC 482, C-104, GJ 74, WR 315, K-850, KWR 108, L-550, BG 212, BG 215, Ghaffa, CPS-1, UC 27, Vardan, Vijay, Vishal, Annigeri, ILWC 7–1, ILWC 33/S-4, CM 368/93, CM 444/92, FLIP 00-17C, FLIP 02-7C, FLIP 02-9C, FLIP 02-40C, FLIP 02-47C, FLIP 03-26C, FLIP 03-29C, FLIP 03-57C, FLIP 03-108C, FLIP 03-127C, FLIP 05–28, FLIP 05-68C, FLIP 05-72C, FLIP 05-85C, FLIP 05-106C, FLIP 90-131C, FLIP 99-66C	Sharma et al. (2005) Sharma and Muehlbauer (2007) Singh et al. (2009) Ali et al. (2011) Kumar et al. (2011)
Botrytis gray mold	ICCV 2, Pusa 209, Gaurav	Singh et al. (2009)
Rust	FLIP05-74C, PI 593072, PI 642748	Rubiales et al. (2001)

5.9 Future Prospects

The cultivated chickpea has limited variability that necessitated using wild *Cicer* species having a high degree of resistance to many biotic and abiotic stresses. Transferring resistance and other desirable gene complexes from such unexploited wild to cultivated species through hybridization are limited by reproductive barriers that can be overcome by using novel biotechnological approaches. Further, a greater understanding of the genetic bases of virulence, mechanism of resistance, and host-pathogen interactions is required to enhance the breeding efficacy in chickpea. Minor diseases have been poorly studied due to difficulty in resistance screening and other reasons which require much more attention in the context of the climate change scenario.

References

Ali H, Ahsanul HM, Shah TM, Rahman M. Validation of molecular markers for resistance among Pakistani chickpea germplasm to races of *Fusarium oxysporum* f. sp. *ciceris*. Eur J Plant Pathol. 2011;132:237–44.

Anonymous. Annual Report, Directorate of Pulses Development, New Delhi; 2016, p. 217.

Arif AG, Jabbar A. A study of physiologic specialisation in Myco sphaere llarabiei Kovaceveski Ascochytarabiei (Pass.) Lab., the causal organism of gram blight. West Pakistan J. Agric Res. 1965;3:103–21.

Bahadur P, Sinha S. Physiologic specialization in *Uromyces ciceris-arietini*. Indian Phytopathol. 1970;23:626–8.

Baite MS, Dubey SC. Race profiling diversity and development of molecular markers for detection of Ascochyta rabiei causing blight in chickpea. Ph.D Thesis; 2015. http:/krishikosh.egranth.ac.in//handle//1510023054.

Bouslama M. Chickpea improvement in Tunisia. In Proceedings of the International Workshop on Chickpea Improvement, 28 February–2 March 1979. International Crops Research Institute for the Semi-Arid Tropic; 1980.

Cunnington JH. Organization of the mitochondrial genome of *Fusarium oxysporum* (anamorphic Hypocreales). Mycoscience. 2007;48:403–6.

Dubey SC, Singh SR. Virulence analysis and oligonucleotide fingerprinting to detect diversity among Indian isolates of *Fusarium oxysporum* f. sp. *ciceris* causing chickpea wilt. Mycopathologia. 2008;165:389–406.

Dubey SC, Priyanka K, Singh V, Singh B. Race profiling and molecular diversity analysis of *Fusarium oxysporum* f. sp. *ciceris* causing wilt in chickpea. J Phytopathol. 2012; https://doi.org/10.1111/j.1439-0434.2012.01954.

Ghosh R, Sharma M, Telangre R, Pande S. Occurrence and distribution of chickpea diseases in central and southern parts of India. Am J Plant Sci. 2013;4:940–4.

Haware MP, Nene YL. Symptomless carriers of the chickpea wilt *Fusarium*. Plant Dis. 1982a;66:809–10.

Haware MP, Nene YL. Races of *Fusarium oxysporum* f. sp. *ciceris*. Plant Dis. 1982b;66:809–10.

Honnareddy N, Dubey SC. Pathogenic and molecular characterization of Indian isolates of *Fusarium oxysporum* f. sp. *ciceris* causing chickpea wilt. Curr Sci. 2006;91:661–6.

ICRISAT. Crop genomics: present and future. In: VI Next Generation Genomics & Integrated Breeding for Crop Improvement Conference, December 06–07, 2017, ICRISAT, Patancheru, Hyderabad, India; 2017.

Iftikhar K, Ilyas MB. Screening of chickpea germplasm against dry root rot disease (*Macrophomina phaseolina*) in pots/glass house. Pakistan J Phytopathol. 2000;12:66–70.
Ilyas MB, Chaudhry MA, Javed N, Ghazanfar MU, Ahsan KM. Sources of resistance in chickpea germplasm against ascochyta blight. Pak J Bot. 2007;39:1843–7.
Jayalakshmi SK, Usharani S, Benagi MDD. Sources of resistance to dry root rot of chickpea caused by *Rhizoctonia bataticola*. Agric Sci Dig. 2008;28:147–8.
Jiménez-Díaz RM, Castillo P, del Mar J-GM, Landa BB, Navas-Cortés JA. Fusarium wilt of chickpeas: biology, ecology and management. Crop Prot. 2015;73:16–27.
Jimenez-Gasco MM, Perez-Artes E, Jimenez-Diaz RM. Identification of pathogenic races 0, 1B/C, 5 and 6 of *Fusarium oxysporum* f. sp. *ciceris* with random amplified polymorphic DNA (RAPD). Eur J Plant Pathol. 2001;107:237–48.
Kaiser WJ, Danesh D, Okhovat M, Mossahebi H. Diseases of pulse crops (edible legumes) in Iran. Plant Dis Rep. 1968;52:687–91.
Kumar J, Choudhary AK, Solanki RK, Pratap A. Towards marker-assisted selection in pulses: a review. Plant Breed. 2011;130:297–313.
Malhotra RS, Khalaf G, Hajjar S, Arslan S. Interspecific hybridization in chickpea. In: Chickpea research for the millennium: Proceedings of the International Chickpea Conference. 20–22 January 2003, Chhattisgarh, India; 2003, pp. 39–49.
Mitra M. Report of the imperial mycologist. Sci Rep Agric Res Inst. 1931;19:58–71.
Nene YL, Shelia VK, Sharma SB. A world list of chickpea and pigeonpea pathogens. 5th ed. Hyderabad: International Crops Research Institute for Semi-Arid Tropics; 1996; p. 27.
Nene YL. A review of Ascochyta blight of chickpea. In: Saxena MC, Singh KB, editors. Proceedings of the Workshop on Ascochyta blight and Winter Sowing of Chickpeas: ICARDA; 1981.
Nene YL, Haware MP. Screening chickpea for resistance to wilt. Plant Dis. 1980;64:379–80.
Pande S, Rao JN, Sharma M. Establishment of the chickpea wilt pathogen *Fusarium oxysporum* f. sp. ciceris in the soil through seed transmission. Plant Pathol J. 2007;23:3–6.
Papavizas GD. Status of applied biological control of soil-borne plant pathogens. Soil Biol Biochem. 1973;5:709–20.
Pratap A, Chaturvedi SK, Tomar R, Rajan N, Malviya N, Thudi M, et al. Marker-assisted introgression of resistance to fusarium wilt race 2 in Pusa 256, an elite cultivar of desi chickpea. Mol Genet Genomics. 2017;292:1237–45.
Rao PA, Haware MP. Inheritance of dry root rot (*Rhizoctonia bataticola*) resistance in chickpea (*Cicer arietinum*). Plant Breed. 1987;98:349–52.
Rubiales D, Moreno I, Moreno MT, Sillero JC. Identification of partial resistance to chickpea rust (*Uromyces ciceris-arietini*). In: AEP (ed). Proc. 4th European Conference on grain legumes, Cracow, Poland; 2001, pp. 194–195.
Sharma KD, Chen W, Muehlbauer FJ. Genetics of chickpea resistance to five races of *Fusarium* wilt and a concise set of race differentials for *Fusarium oxysporum* f. sp. *ciceris*. Plant Dis. 2005;89:385–90.
Sharma KD, Muehlbauer FJ. *Fusarium* wilt of chickpea: physiological specialization, genetics of resistance and resistance gene tagging. Euphytica. 2007;157:1–14.
Sharma M, Ghosh R, Pande S. Dry root rot (*Rhizoctonia bataticola* (Taub.) Butler): an emerging disease of chickpea – where do we stand? Arch Phytopathol Plant Protect. 2016;48:798–812.
Sharma M, Ghosh R. An update on genetic resistance of chickpea to Ascochyta blight. Agronomy. 2016;6:1–15.
Sharma M, Pande S. Unravelling effects of temperature and soil moisture stress response on development of dry root rot [*Rhizoctonia bataticola (Taub.)*] butler in chickpea. Am J Plant Sci. 2013;4:584–9.
Sharma M, Ghosh R, Krishnan RR, Nagamangala UN, Chamarthi SK, Varshney R, et al. Molecular and morphological diversity in Rhizoctonia bataticola isolates causing dry root rot in chickpea (*Cicer arietinum* L.) in India. Afr J Biotechnol. 2012;11:8948–59.
Sharma OP, Gupta RBL. Fungicides in the control of chickpea dry root rot caused by *Rhizoctonia bataticola*. J Mycol Plant Pathol. 2004;34:321–2.

Sharma YK, Gaur RB, Bisnoi HR. Cultural, morphological and physiological variability in *Macrophomina phaseolina*. J Mycol Plant Pathol. 2004;34:532–4.
Shivanand N, Jagadeesh S, Pavan J, Shilpa M, Kanakaraddi SG. Analysis and estimation of rust disease in Bengal gram based on thresholding and RGB extraction using image processing. IJARCET. 2014;3:1502–5.
Sillero J, Moreno-Alías I, Rubiales D. Identification and characterization of resistance to rust (*Uromyces ciceris-arietini* (Grognot) Jacz. & Boyd) in a germplasm collection of *Cicer* spp. Euphytica. 2012;188:229–38.
Singh KB, Dahiya BS. September Breeding for wilt resistance in chickpea. In: Symposium on Wilt Problem and Breeding for Wilt Resistance in Bengal Gram. Indian Research Institute, New Delhi, India; 1973. pp. 13–14.
Singh KB, Reddy MV, Nene YL. International testing of chickpeas for resistance to Ascochyta blight. Plant Dis. 1984;68:782–4.
Singh NP, Sewak Shiv, Iquebal MA, Chaturvedi SK, Nath O. Improved varieties of chickpea in India. Technical Bulletin, IIPR, Kanpur; 2009.
Singh PJ, Mehrotra RS. Penetration and invasion of gram roots by *Rhizoctonia bataticola*. Indian Phytopathol. 1982;35:336–8.
Singh SK, Nene YL, Reddy MV. Some histopathological observations of chickpea roots infected by *Rhizoctonia bataticola*. Int Chick News. 1990;23:24–5.
Smithson JB, Thompson JA, Summerfield RJ. Chickpea (*Cicer arietinum* L.). In: Summerfield RJ, Roberts EH, editors. Grain legume crops. London, UK.: Collins; 1985. p. 312–90.
Thaware DS, Kohire OD, Gholve VM. Integrated management of wilt of chickpea (*Cicer arietinum* L.) caused by *Fusarium oxysporum* f. sp. ciceris. Green Farming. 2016;7:687–93.
Thilagavathi R, Saravanakumar D, Ragupathi N, Samiyappan R. A combination of biocontrol agents improves the management of dry root rot (*Macrophomina phaseolina*) in greengram. Phytopathol Mediterr. 2007;46:157–67.
Trapero-Casas A, Jiménez-Díaz RM. Fungal wilt and root rot diseases of chickpea in southern Spain. Phytopathology. 1985;75:1146–51.
Tripathi HS, Singh RS, Chaube HS. Effect, of sun drying on the recovery of A. rabiei from infected chickpea seeds.Int. Chickpea Newsl. 1987;16:13–14.
Varshney RK, Nayak SN, May GD, Jackson SA. Next generation sequencing technologies and their implications for crop genetics and breeding. Trends Biotech. 2010;27:522–30.
Varshney RK, Mohan SM, Gaur PM, Gangarao NV, Pandey MK, Bohra A, et al. Achievements and prospects of genomics-assisted breeding in three legume crops of the semi-arid tropics. Biotechnol Adv. 2013a;31:1120–34.
Varshney RK, Song C, Saxena RK, Azam S, Yu S, Sharpe AG, et al. Draft genome sequence of chickpea (*Cicer arietinum*) provides a resource for trait improvement. Nat Biotechnol. 2013b;31:240–6.
Vidhyasekaran P, Muthamilan M. Development of formulations of *Pseudomonas fluorescens* for control of chickpea wilt. Plant Dis. 1995;79:782–6.
Vir S, Grewal JS, Gupta VP. Inheritance of resistance to *Ascochyta* blight in chickpea. Euphytica. 1975;24:209–11.
Warda J, Mariem B, Amal B, Mohamed B, Mohamed K. *Fusarium* Wilt Affecting Chickpea Crop., Agriculture. 2017;7:23.
Westerlund FV, Jr-Cambell RN, Kimble KA. Fungal root rots and wilt of chickpea in California. Phytopathology. 1974;664:432–6.
Zhu C, Gore M, Buckler ES, Yu J. Status and prospects of association mapping in plants. Plant Genom. 2008;1:5–20.

Chapter 6
Wilt and Root Rot Complex of Important Pulse Crops: Their Detection and Integrated Management

Nidhi Kumari and Shabnam Katoch

6.1 Introduction

Attainment of self-sufficiency in crop production is the only way to fulfil the food requirement of expanding population; though India has come a long way from a pulse-deficient country to self-reliant one, still there are so many factors which contribute towards low production of agricultural goods. Among different cultivated crops, pulses are in the midst of imperative sources which have a say to the nutritional security of a country (Singh et al. 2015). Pulses are protein-rich commodity which in addition to the fulfilment of protein requirement also improves the soil fertility (Narayan and Kumar 2015; Singh et al. 2018). Throughout the world, India has the biggest contribution in the production and consumption of tropical and sub-tropical pulse crops such as gram, red gram, black gram, green gram, field pea and lentil (Srivastavaa et al. 2010; Singh et al. 2017a; Hasan and Khan 2018). In India, pulses are cultivated in 294.65 Lakh hectares of land with the annual production of 22.95 MT (http://agricoop.nic.in/sites/default/files/Krishi%20AR%20 2017-18-1%20for%20web.pdf), out of which the maximum share of 77.0% is collectively from Madhya Pradesh, Maharashtra, Uttar Pradesh, Karnataka, Andhra Pradesh and Rajasthan followed by only 23% from Gujarat, Chhattisgarh, Bihar, Orissa and Jharkhand (Trivedi et al. 2017; Singh et al. 2018). In spite of country's autonomy in pulse production, there are some supply and demand affecting ambiguities like unpromising weather, several agronomic limitations, insect-pests and diseases and inappropriate marketing. Out of all, the effect of microbes on plant growth and development has arrived as a major apprehension among the pulse

N. Kumari (✉)
Division of Crop Protection, ICAR-CISH, Lucknow, Uttar Pradesh, India

S. Katoch
Department of Plant Pathology, CSK HPKV, Palampur, Himachal Pradesh, India

B. P. Singh et al. (eds.), *Management of Fungal Pathogens in Pulses*, Fungal Biology, https://doi.org/10.1007/978-3-030-35947-8_6

growers of India. Out of ~100 fungal pathogens, the soilborne disease incidents causing wilts and root rots have been a matter of worry for sustainable production of pulses since couple of years and are reported to cause considerable yield losses in pulses (Trivedi et al. 2017), sometimes even up to 100% (Sinha et al. 2018). The complex nature of these diseases further aggravates the problem. The vascular wilt caused by *Fusarium* sp. is one of the major threats to pulse growers throughout the world (Sinha et al. 2018). Keeping in view the importance of pulses, different management strategies, viz. cultural, physical, biological and chemical control methods, have been used to manage wilt and root rot diseases/complex, but till date, apart from the high cost and deleterious effects of chemicals, their use is considered as the quick and accurate way of disease management. But as far as soilborne pathogens are concerned, the sole use of fungicides may not lead to their proper management. And due to the increasing awareness regarding health hazards caused by the intake of food with pesticide residues, scientists are looking for integrated management strategies which are not solely dependent on the use of chemicals. Before the implementation of any management strategy, the accurate detection of diseases is also very important. The early detection of diseases at their onset helps farm scientists and farmers to plan as well as execute effective integrated management strategies. From the last many years, the conventional methods are in use for pathogens detection, but these methods sometimes lead to confusing conclusions and are not as accurate as DNA and protein-based methods (Katoch et al. 2019). In this chapter, we will discuss the different wilts and root rots (Table 6.1) causing considerable losses to important Indian pulse crops and their integrated/holistic management. In addition to this, recent diagnostic methods used for early and timely detection will also be discussed.

6.2 Root Rot of Pulse Crops

Root rots are the diseases of utmost importance impacting a wide range of crops worldwide. Often root rot is a complex disease where more than one pathogen is involved. Fungi, oomycetes, bacteria and viruses have been reported to cause root rots (El Karkouri et al. 2010; Legg et al. 2011; Heffer Link et al. 2002; Cleary et al. 2011; Cui et al. 2014). Frequently nematodes have been reported to aggravate the problem by facilitating the entry of other pathogens through wounds made by nematodes while feeding (Back et al. 2002). Initial symptoms appear on the roots of the affected plants which go unnoticed or are not visible. Till the aboveground symptoms become noticeable, sufficient losses to the plant health have already occurred; thus it becomes almost impossible to recover the plants. Root rot is favoured by poor drainage conditions, moderate to high soil moisture, monocropping, etc.

Table 6.1 List of important wilt and root rots of major pulse crops in India

Crop	Disease	Causal organism	References
Gram	Dry root rot	*Rhizoctonia bataticola (taub)* Butl. (Pycnidial stage: *Macrophomina phaseolina* Tassi Goid)	Pandey et al. (2017); Kadam et al. (2018); Sunkad et al. (2018)
	Collar rot	*Sclerotium rolfsii* (Teleomorph: *Athelia rolfsii* (Curzi) Tu and Kimbrough)	Ghosh et al. (2013); Ahsan et al. (2018)
	Black root rot	*Fusarium solani*	Ghosh et al. (2013)
	Wilt	*Fusarium oxysporum* f. sp. *ciceris*	Pandey et al. (2017); Sankar et al. (2018)
Red gram	Dry root rot	*Rhizoctonia bataticola* (Taub.) Butler (*Macrophomina phaseolina* (Tassi) Goid)	Maruti et al. (2017)
	Wilt	*Fusarium udum* Butler	Chennakesavulu et al. (2013); Singh et al. (2016); Saxena et al. (2012); Sharma et al. (2018)
Black gram	Dry root rot	*Rhizoctonia bataticola* (Taub.) Butler (*Macrophomina phaseolina* (Tassi) Goid)	Tetali et al. (2015)
Green gram	Root rot	*Rhizoctonia bataticola* (Taub.) Butler (*Macrophomina phaseolina* (Tassi) Goid)	Sarkar and Bhattacharyya (2008); Mallaiah and Rao (2016); Shahid and Khan (2016)
	Seedling rot and web/ leaf blight	*Rhizoctonia solani* Kuhn (*Thanatephorus cucumeris*)	Singh et al. (2013)
Field pea	Rhizoctonia root rot	*Rhizoctonia solani*	Rawat et al. (2014)
	Wilt/root rot complex	*Fusarium oxysporum* f. sp. *pisi* *Fusarium solani* f. sp. *pisi* (Jones) Synder and Hansen	Rao (2014); Thakur et al. (2016); Nongmaithem et al. (2017)
Lentil	Collar rot or root rot	*Sclerotiun rolfsii (Athelia rolfsii)*	Surulirajan et al. (2007); Kushwaha (2016); Tiwari et al. (2018)
	Rhizoctonia root rot	*Rhizoctonia solani*	Tiwari et al. (2018)
	Wilt	*Fusarium oxysporum* f. sp. *lentis*	Garkoti et al. (2013); Singh et al. (2017a, b); Arya and Kushwaha (2018)

6.2.1 Dry Root Rot of Chickpea

Chickpea (*Cicer arietinum* L.) is grown over an area of 9.53 million hectare with 9.07 million tonnes in India (FAOSTAT 2017). Dry root rot (DRR) of chickpea caused by *Rhizoctonia bataticola* (Taub.) Butler. (synonym: *Macrophomina phaseolina*) has emerged as a serious problem of world's second largest and India's largest produced pulse crop (Sharma et al. 2015). The life cycle of *Rhizoctonia*

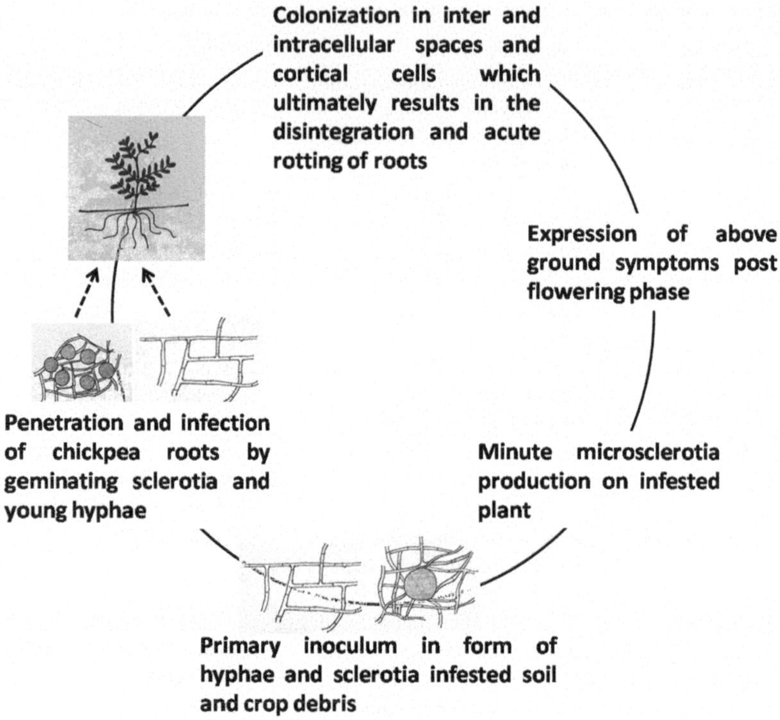

Fig. 6.1 Disease cycle of dry root rot of chickpea caused by *Rhizoctonia bataticola*

bataticola is being illustrated in Fig. 6.1. As per recent reports after *Fusarium* wilt (*Fusarium oxysporum* f. sp. *ciceris*), DRR is imposing humongous hazard on chickpea production worldwide (Ghosh et al. 2013; Sharma et al. 2015).

Pathogen:	Taxonomic position of *Rhizoctonia bataticola* (Taub.) Butler
Kingdom:	Fungi
Division:	*Basidiomycota*
Class:	*Agaricomycetes*
Order:	*Cantharellales*
Family:	*Ceratobasidiaceae*
Genus:	*Rhizoctonia*

Augustin Pyramus de Candolle descried the fungus *Rhizoctonia* (means "root killer") in 1815 as plant pathogenic fungi capable of producing hyphae and sclerotia. The fungus is predominantly saprophytic in nature but acts as facultative parasite causing diseases to many economically important crops (Ram and Singh 2018). Though the accurate taxonomic name recognized is *M. phaseolina* (CMI description of pathogenic fungi and bacteria No.275), *R. bataticola* is referred for the sclerotial phase of the fungus (Holliday and Punithalingam 1970). *R. bataticola* is a serious soilborne pathogen capable of infecting greater than 500 cultivated and wild host plants (Maruti et al. 2017). In DRR of chickpea, only sclerotial phase is present; therefore the pathogen is referred as *R. bataticola.*

Symptoms: Symptoms are generally not visible at seedling stage. Older plants are more prone to the disease (Sharma and Pande 2013). Symptoms are more evident during post-flowering period as chlorosis of petioles and leaflets followed by drooping at the top of the plant. The leaves and stem become straw coloured, while few times the lower stem and leaves become brown coloured (Sharma et al. 2015; Ram and Singh 2018). Upon uprooting the diseased plant, blackened and rotten tap roots with fewer or no lateral and finer roots are observed. These dead roots become brittle with shredded bark. Microsclerotia can be clearly seen underneath the bark.

Disease cycle: The primary inoculum remains in soil in the form of hyphae and sclerotia. The epidermal cells are dismantled by the enzymatic actions and mechanical pressure exerted by the pathogen followed by penetration of roots, though the infection may also take place during emergence of seedlings through cotyledons or through wounds on root surface and small rootlets. Mechanical plugging of xylem vessels due to microsclerotia and toxin production also takes place during disease development along with the secretion of macerating enzymes (Bhatt and Vadhera 1997; Sharma et al. 2004). After penetration the hyphae grow inter- and intracellularly and spread through the cortical cells which ultimately results in the disintegration and acute rotting of roots (Singh and Mehrotra 1982). The colonization of vascular system by hyphae and plugging of xylem vessels by sclerotia is observed in this disease (Singh et al. 1990). As the disease advances, the root necrosis constantly extends without any evident aboveground symptoms till flowering and podding stage (Fig. 6.1).

6.2.2 *Dry Root Rot of Pigeon Pea*

Red gram (pigeon pea, *Cajanus cajan* (L.) Millsp., $2n = 22$), after chickpea, is the second predominant pulse crop in India and can be cultivated in low fertilizer input land or even in drought conditions. This pulse crop is famous among the small and marginal farmers due to its hardy, wide adapting and drought-tolerating nature. In India the cultivation of this pulse crop is expanded over 5.38 million hectare land with 4.87 million tonnes production (FAOSTAT 2017). Among various constraints in achieving maximum productivity of pigeon pea in India, one is dry root rot of pigeon pea which is distributed in Uttar Pradesh, Madhya Pradesh, Karnataka, Maharashtra, Tamil Nadu and Delhi states of the country. Disease comes in severe form in late-sown or summer pigeon pea as well as in perennial or rationed pigeon pea. Under favourable conditions the disease may result in 100% yield loss (Smitha et al. 2015).

Causal organism: Sclerotial stage, *R. bataticola*; pycnidial stage, *M. phaseolina*

Symptoms: There is drooping and drying of leaves followed by sudden drying and death of the plants. During early disease stage, on stems and branches, spindle-shaped lesions surrounded by brown margins with grey centres and pycnidial bodies scattered all over are formed which later on coalesce resulting in drying and ultimately death of the branches or even the whole plant. The infected plants have rotten, shredded and brittle roots. Underneath the bark of finer roots, dark, blackened streaks with dark sclerotial bodies are quite evident. Prolonged hot and dry weather or drought followed by irrigation and rains favours the disease development.

6.2.3 Root Rot and Leaf Blight of Black Gram (*Vigna mungo*) and Green Gram (*Vigna radiata*)

Black gram and green gram belong to Fabaceae and are widely cultivated in Indian subcontinent as sole, mixed, catch or sequential crop in kharif or summer season under rainfed or semiarid conditions. In India, black gram is popularly known as "urad dal" and is one of the highly prized pulse crops in India. In India, it is consumed in the form of dal (husked or non-husked, whole or split). In Indian subcontinent, green gram/golden gram also called as mung or mung bean is widely cultivated as short-duration pulse crop grown in kharif, summer and spring seasons. Root rot and leaf blight also called as web blight is one of the major constraints in the production of black and green gram. The distribution of the disease is widespread and has been reported from India, Malaysia, the Philippines, Iran and Taiwan.

Causal organism: *Rhizoctonia solani* {Perfect stage: *Thanatephorus cucumeris* (Frank) Donk}

The fungus is omnipresent and can be easily isolated from infected plant part and soil. The fungus has characteristic septate mycelium, white to deep brown in colour with right angled branching.

Symptoms: During initial phase of the disease, the symptoms are damping off, seed and root rot, seedling blight, stem canker and web blight. On seedling hypocotyls, reddish brown sunken lesions which later on enlarge and coalesce lead to girdling of stems which ultimately results in death of the affected seedlings.

Thanatephorus cucumeris (Frank) Donk causes the web blight symptoms on the foliage of black and green gram. The symptoms include yellowing of leaves followed by appearance of brown irregular lesions initiating from the apical portion of the leaflets later on covering the entire leaf blades and then advances to the petiole and stem part. The fungal runner hyphae can be seen on affected leaves, petioles and stem, thus causing the typical web blight symptoms. Under severe infection, the affected plant die prematurely even before the commencement of flowering stage. Fewer numbers of pods with brown necrotic lesions on their surface are produced by the infected plants. As the disease advances, on affected plant parts and fallen leaves, an abundant number of sclerotia which are initially white in colour but later on turn brown are formed.

Disease cycle: The fungus grows saprophytically in the soil enriched with adequate amount of organic matter. The wide host range and regular addition of organic matter in the soil allows the survival of pathogen in soil for longer duration. Sclerotia produced by fungi persist in soil, and its germination is stimulated by the root exudates of the host plants under favourable humidity and temperature conditions. The soil inoculum is disseminated by flooding, irrigation, movement of contaminated soil and plant debris. Basidiospores are produced by *T. cucumeris* on healthy areas adjoining the infected part which cause the aerial infection on plants. Temperature around 20 °C as well as wet and alkaline soil favours the rapid disease development.

6.2.4 Root Rot and Damping Off of Cowpea

Cowpea [*Vigna unguiculata* (L.) walp] is mainly cultivated in northern and central part of India as annual leguminous fodder crop. Cowpea is susceptible to many insects, bacteria, fungi and viruses that are capable of infecting at all growth stages of the crop. The root rot and damping off of cowpea caused by *R. solani*, *M. phaseolina* and *Pythium ultimum* are the most devastating disease occurring as a complex.

Symptoms: The disease is mainly characterized by rapid death of young plants. The other symptoms include yellowing and drying of leaves, rooting of taproots with longitudinal cracks on stems which ultimately results in poor yields.

6.2.5 Charcoal Rot, Ashy or Stem Blight or Dry Root Rot of Soybean

Soybean (*Glycine max* L. Merril) is an important oilseed crop contributing about 25% of global edible oil (Agarwal et al. 2013). The USA, Argentina and Brazil occupy top three positions of leading soybean producers in world. In India soybean has been introduced by China and now is being cultivated on an area of 10.60 million hectare with 10.98 million tonnes production (FAOSTAT 2017) in Madhya Pradesh, Maharashtra and Rajasthan which together contribute for more than 90% of total production from the country. Charcoal rot, also known as DRR, dry weather wilt, ashy stem blight and seedling blight disease, caused by *M. phaseolina* (Tassi) Goid is one of the major diseases of soybean (Su et al. 2001). In India the disease was of minor importance till 2004 but acquired the status of major disease due to altered weather conditions (Agarwal et al. 2013).

Causal organism: *M. phaseolina* (Tassi) Goid

The pathogen has a wide host range which includes major field and pulse crops like common bean, soybean, mungbean, cotton, maize, sorghum, sesame, peanut, cowpea and chickpea (Dhingra and Sinclair 1977; Diourte et al. 1995).

Symptoms: The pathogen is predominant soilborne pathogen but also seed borne in nature, capable of infecting the crop at any growth stage. The symptoms on cotyledons appear as dark brown spots after emergence with brown to black margins, and they shed at an early stage. After the emergence of unifoliate leaf, reddish brown, circular to oblong lesions which after several days may turn dark brown to black appear on the emerging hypocotyls of infected seedlings. Lesions appear on roots, stems, pods and seeds. Lower leaves become chlorotic and later on wilt and dry. As the disease advances, reddish brown discolouration of the vascular elements of roots and lower stems followed by premature yellowing of plants is observed. Blackening and cracking of roots is the most common symptom of this disease. Diseased plants show poor seed-setting in pods with reduced seed size, which ultimately lead to heavy yield losses.

Disease cycle: Pin-sized microsclerotia are produced in abundance underneath the epidermal tissue of the affected lower stems and roots after the death of plants. These microsclerotia are capable of long survival up to 2–12 years in soil and initiate the disease by acting as primary inoculum (Meyer 1974). The pathogen has also been associated with seed when detected using agar plate, blotter paper and modified potato-sucrose-agar [PSA + Penta Chloro Nitro Benzene (PCNB)] methods by Kushi and Khare (1978). The germination of microsclerotia is induced by the root exudates of host plants present in the vicinity. Heavily infected plants die at early stage due to the accumulation of fungal toxin, viz. botryodiplodin or phaseolinone (Ramezani et al. 2007). Microsclerotia are released into the soil after the death of the plant and the cycle continues.

6.3 Wilt Diseases of Major Pulse Crops

6.3.1 Fusarium Wilt of Chickpea

Butler (1918) first time reported the occurrence of gram wilt from India, and later in 1940, Padwick identified *Fusarium orthoceras* var. *ciceri* as the incitant of chickpea wilt (Chand and Khirbat 2009). Due to the complex nature of gram wilt, Dastur (1935) came to the conclusion that the drooping of plants was because of *Rhizoctonia* wilt caused by *Rhizoctonia bataticola*. In 1940, Synder and Hansen renamed *Fusarium orthoceras* var. *ciceri* as *F. oxysporum* f. sp. *ciceri*, and it is now worldwide accepted. All over the world, gram wilt alone is known to cause 10–50% yield losses, while from India 10–15% losses were observed (Kheni et al. 2017).

Symptoms: Conducive conditions to *Fusarium* wilt pathogen initially results in drooping, yellowing and drying of the leaves followed by the wilting of entire plant (Lodhi et al. 2006; Kumari and Khanna 2018). Most of the times, disease appears in scattered patches of yellow colour, but under favourable conditions, wilting of entire field may occur. Sometimes, the infection starts after 25 days of sowing and that diseased condition is known as early wilt. In early wilt, the seedlings lose their turgor, further collapse and lie flat on the field. In most of the cases, the prominent disease symptoms appear at 6–8 weeks after sowing when flowering starts and during pod formation stage; that situation is known as late wilt (Jimenez-Díaz et al. 2015; Arunodhayam et al. 2014). *F. oxysporum* usually results in discolouration, desiccation and collapse/crumpling of entire plant following the drooping of leaves. The drooping starts from the upper portion of the plant, and within no time, entire plant becomes wilted. The cross-sectioned or vertically splitted roots/stems of infected plants shows dark brown discolouration of xylem vessels. The pathogen results in the development of histological distortions of vascular tissues along with the formation of occlusions and gel in the xylem cells (Patil et al. 2017). These histological distortions lead to the clogging of vascular tissues and retard the vascular flow of water, and ultimately, the affected plant wilts. Actually, the toxins produced by *F. oxysporum* f. sp. *ciceri* are responsible for wilting of plants (Chand and Khirbat 2009).

Causal organism: *Fusarium oxysporum* Schlecht and Emnd Snyd. & Hans. f. sp. *ciceri* (Padwick) Snyd. & Hans

F. oxysporum f. sp. *ciceris* produces three types of asexual spores, i.e. macro- and microconidia and chlamydospores. Under in vitro conditions, white mycelial growth with different pigments, viz. pink, pale yellow, light yellow, etc. has been observed (Patra and Biswas 2016). The macroconidia are 25.00–55.00 μm × 2.50–6.00 μm in size, straight to slightly curved, thin walled usually with 3–5 septa, a foot-shaped basal cell and a tapered and curved apical cell, while the microconidia are 5.00–15.00 μm × 2.00–5.00 μm, ellipsoidal with single or no septum (Nath et al. 2017). The chlamydospores are thick walled, globose, formed singly and in pairs or in chains on hyphae or alternatively by the modification of hyphal cells and are important source of primary infection. In the absence of host plant, chlamydospores of the fungus can survive up to 6 years in the soil. In lab, the chlamydospore formation has been observed by several workers in 15-day-old culture and infected tissues. The teleomorph or sexual reproductive stage of *F. oxysporum* is unknown. The variability among different isolates could be studied morphologically by using the size and shape of asexual spores (Sinha et al. 2018). Its growth is primarily dependent on the type of soil, pH, moisture content and temperature. The optimum temperature for disease development is 25 °C, but fungus can grow within a range of 7–35 °C in the soils having pH 4–9.4. Many researchers reported different pH, i.e. 5.1–5.9 and 7.1–7.9 for mycelial growth and sporulation, respectively (Jendoubi et al. 2017).

Disease cycle: The pathogen survives as chlamydospores or mycelium on seed, soil (for up to 6 years) and crop residues buried in the soil (Lodhi et al. 2006). Initial infection starts with the germination of chlamydospores/mycelia after getting stimulus (phytoalexins and flavonoids) from the roots of host/non-host plants. After germination, the germtube directly or through wounds invades the roots and enters into the epidermal cells of the plant. Following penetration, the fungus starts colonizing the root cortex intracellularly and eventually grows to clog xylem vessels. Ultimately *F. oxysporum* results in wilting of plants, and it is dependent on pathogen activities like mycelium formation in vessels; toxin production; production of gels, gums and tyloses; etc. Infested soil serves as the main source of inoculum, and in fallow soils, some dicot weeds are reported in the survival of pathogen (Jendoubi et al. 2017). In the cropping season, mycelium or spores of fungus dispersed in the soil to small extent and cause infection in surrounding plants, in addition to this the inoculum disseminates to distant places through field equipments/seeds/cuttings etc.

6.3.2 Wilt of Pigeon Pea

Pigeon pea wilt is one of the devastating diseases causing even 100% yield losses under favourable disease conditions (Pande et al. 2013). In addition to predisposing factors, the losses are also dependent upon the stage of plant at which pathogen establish itself. There are reports where 30%, 67% and even 100% losses are recorded when infection occurs at preharvesting, maturity and pre-podding stage, respectively. The disease was first time reported by Butler (1906) from Bihar, India, and later in 1910, the causal organism was named as *Fusarium udum*. Thereafter, Rai and Upadhyay (1982) reported its perfect stage *Gibberella indica* Rai and found that perfect stage formed on exposed roots and collar region up to the height of 35 cm.

Symptoms: The prominent symptoms are drooping of plants due to turgidity loss and clogging of xylem vessels. Infected plants show partial wilting, mild interveinal chlorosis, discolouration of xylem vessels and purplish bands on the stem which extends in the upward direction. In addition to this, drying of plants from top towards base following the yellowing and chlorosis is also common. Generally, the wilting of plants is due to the presence of mycelial clumps in the xylem vessels (Chaudhary 2016; Meena 2016). The cross-sections of the main root and base of the stem show tissue discolouration, and in case of partial wilting, plant tissues are discoloured and show wilting from one side, while the rest of the plant escapes.

Causal organism: *Fusarium udum* Butler

The pathogen is host specific and soilborne in nature and can survive on the crop debris for 3 years. Like *F. oxysporum*, it also produces macro- and microconidia and chlamydospores, but presence of prominent apical hook cell of macroconidia makes it different from *F. oxysporum*. The fungus produces septate hyaline mycelium, which grows inter- and intracellularly in xylem vessels to obstruct the water flow. The microconidia formed by *F. udum* are unicellular, 1 or 2 septate and small in size which varies from 5 to 15 × 2 to 4 μm, while the macroconidia are long, and slightly curved, 3–4 septate and 15–50 × 3–5 μm in size. In addition to these, persisting/perennating structures, i.e. chlamydospores, are also formed which are formed at either terminal ends or intercalary regions from any cell of hyphae or from macroconidium. The optimum temperature for its growth and development ranges between 17 and 29 °C with soil pH of 4.6–9.0.

Disease cycle: *Fusarium* wilt of red gram is soilborne in nature, but there exist few reports which confirm its survival in seed. Wilt pathogen has two phases in its life cycle, i.e. pathogenic and saprophytic. In first phase pathogen remains attached to the host plant, while in the later, it survives on dead host plants/parts as conidia or mainly as the chlamydospores. Generally, infected seeds, soil and roots of previous year crop serve as the main source of inoculum. Conidia/chlamydospores germinate and penetrate the rootlets of pigeon pea. After that, the fungal mycelium grows in inter- and intracellular spaces and ultimately clogs the water-conducting vessels of host plant. Following the clogging of xylem vessels, symptoms appear on the infected plant, and on the infected portion, macro- and microconidia and chlamydospores are formed which serve as a both primary and secondary inoculum. The fungus produces different pectic enzymes (pectin methyl esterase, polygalacturonase and cellulase) and toxins (Fusaric acid) which are involved in the pathogenesis.

6.3.3 Wilt of Field Pea

The production and productivity of pea is adversely affected by number of plant pathogens, but *Fusarium* wilt and root rot diseases are of considerable importance (Sharma 2011). The wilt disease for the first time was reported by Jones and Linford (1925) from the USA. Linford (1928), in the initial years of discovery, named the suspected wilt causing entity as *F. othoceras* App. and *Wr.* var. *pisi*. Thereafter in 1935, the pathogen was renamed as race 1 of *F oxysporum* Schlecht f. sp. *pisi* (van Hall) Snyd and Hans.

Symptoms: In pea generally two forms of wilt are known, i.e. wilt and near wilt. Like other wilts, infected plants show drooping and change in colour of foliage from green to pale yellow, and ultimately due to loss of turgidity, the entire plant topples down. Tissue discolouration is also very common and could be observed after cross-sectioning of root or the lower stem. However, in case of near wilt (race 2), the disease appears later at late blossom stage or at pre-pod or full pod development stage. Near wilt is different from normal wilt, as it appears in scattered manner in the field rather than being concentrated in specific areas as in case of race

Causal organism: *Fusarium oxysporum* f. sp. *pisi* (Linford) Snyder & Hansen

The formation of macro- and microconidia and chlamydospores is common like other wilts. In this case the microconidia are oval to cylindrical and 5–12 × 2.2–3.5 μm in size, while macroconidia are 3–5 septate, fusoid, pointed at both of the ends and 27–46 × 3–4.5 μm in size. Chlamydospores are also formed.

Disease cycle: Most of the *Fusarium* causing wilt diseases are soilborne in nature and survive for a longer period in a soil by means of chlamydospores. *F. oxysporum* f. sp. *pisi* is reported to remain viable for more than 10 years, and once the pea crop is available in the field, the pathogen penetrates into the rootlets and ultimately

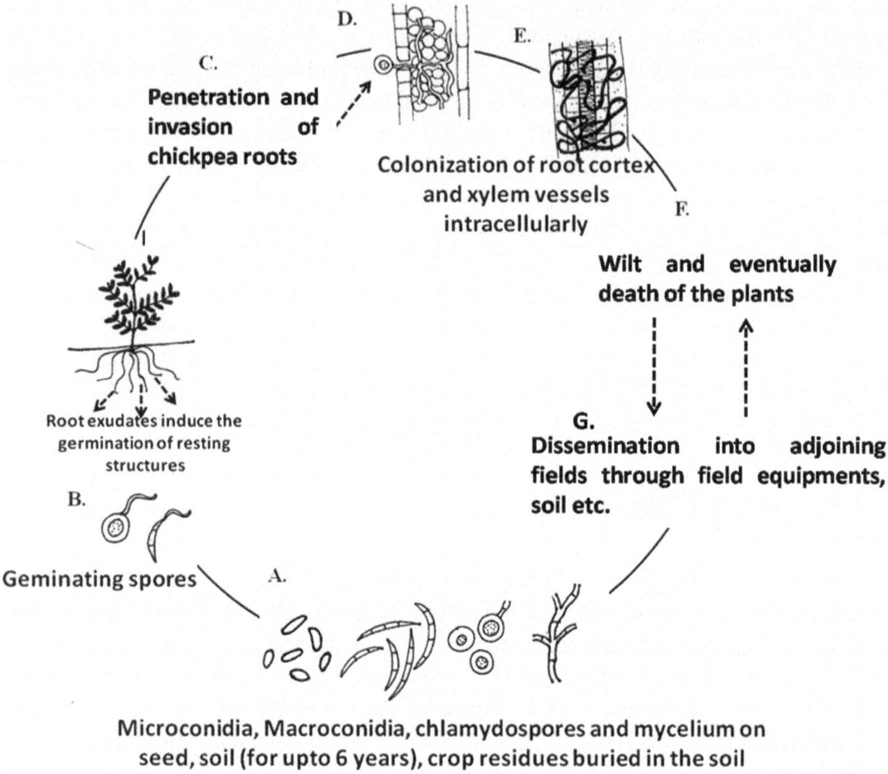

Fig. 6.2 Disease cycle of *Fusarium* wilt of chickpea caused by *Fusarium oxysporum* f. sp. *ciceris*

enters into the vascular system of the plant. The seed-borne nature of the pea wilt pathogen has also been reported. After entering into the roots or cortex, the pathogen established it in the xylem vessels, and further pathogen enters into the system of plants and results in seed-borne infections. The fungus continues to grow on the crop debris left after the death of plants and resulting in the establishment of soil-borne inoculum. The fungus is monocyclic in nature (Fig. 6.2).

6.3.4 Wilt of Lentil

In India, *Fusarium* wilt is a major constraint behind low production of lentil, and from 50% to complete yield losses are reported under favourable conditions (Tiwari et al. 2018). The severity of lentil wilt is dependent on different factors including crop stages, predisposing factors and variety sown in the field. Chaudhary et al. (2009) reported the association of three fungal pathogens *Fusarium oxysporum* f.

sp. *lentis*, *Sclerotium rolfsii* and *Rhizoctonia bataticola* with the wilt/root rot complex from Indian conditions and found the dominance of wilt pathogen.

Symptoms: Like other wilts, lentil wilt can be prevalent at seedling and adult plant stages. The infection at seedling stage leads to the drooping and toppling of lentil seedlings, and the condition is referred to as early wilting. At this stage the roots appear healthy and no tissue discoloration is observed, while the infection at adult stages of plant, i.e. flowering to pre-podding stage, result*s* in either partial or complete wilting of infected plants. Flowering to pre-podding stage is considered as the crucial stage, and infection at these stages leads to complete crop l**o**ss. The optimum temperature for pathogen ranges between 22 and 25 °C (Tiwari et al. 2018). The infection at later stages is characterized by dull green foliage, sudden drooping of top leaves and branches followed by wilting of the entire plant.

Causal organism: *Fusarium oxysporum* Schlecht. emend Snyder & Hansen f.sp. *lentis* Vasudeva and Srinivasan

Lentil wilt causing entity, i.e. *Fusarium oxysporum* f. sp. *lenis*, was first time reported by Booth in 1971. Similar to other *Fusarium* spp., the fungal mycelium is septate, and all the three asexual spores are formed in *F. o* f. sp. *lentis*. The microconidia are straight or curved and 5–11 × 2.5–3.5 μm in size while, macroconidia are fusoid, 1–6 septate and 25–65 × 3.5–4.5 μm. The chlamydospores' formation under in vitro conditions (on old cultures) has also been observed. The host range studies by many workers on different crops concluded that the *Fusarium oxysporum* f. sp. *lentis* produces disease only on lentil.

Disease cycle: The pathogen is soilborne and known to survive in soil for 3–4 years without its host. The primary infection is through chlamydospores which remain viable for the next season or for longer periods. Secondary spread is through conidia by irrigation water, cultural operations and implements.

6.4 Recent Advances in Detection and Diagnosis of Plant Diseases

Fungi are the most diverse plant pathogens with a wide host range accounting for 70–80% of diseases infecting field crops, vegetables, fruit trees and ornamental plants (Ray et al. 2016). Till date fungal disease management is still a challenge due to wrong diagnosis of disease, resistance breakdown in host plants, development of fungicide resistance in pathogens, residual effect of fungicide in environment, etc. Soil has a complex environment, thus forge myriad of challenges in detection, isolation and quantification of soilborne pathogens. Timely and accurate disease detection in case of soilborne pathogens in the absence of their hosts has always remained a limitation (DeShields et al. 2018). The soilborne pathogens infect the plants resulting in early symptomless infection phase and express the symptoms when sufficient impact on plant yield and productivity has already taken place. Accurate disease detection and diagnosis to clearly define the plant disease and causal agent is the first and foremost step in the integrated disease management. This allows preventing the introduction and establishment of a soilborne pathogen to newer areas and losses due to planting of healthy planting material in already infested soil, restricting the movement of infested soil and water to a possible extent. The current detection methods include conventional and advanced molecular methods (Fang and Ramasamy 2015; Balodi et al. 2017). Conventional methods include identification based on diseased symptoms, isolation and culturing of the pathogen on artificial regular or selective media followed by microscopic observations, growing of healthy plants on soil under test, etc. But all these methods are time-consuming and laborious, require skilled laboratory staff and often lead to incorrect diagnosis or wrong interpretation (Tsedaley 2015). Molecular-based approaches are competent strategies in case of early-stage detection and are very helpful in undertaking prophylactic measures.

6.5 Molecular Approaches in Plant Disease Diagnosis

6.5.1 Polymerase Chain Reaction (PCR)

PCR method that involves in vitro replication of DNA was first invented by Kary Mullis in 1984 for which in 1993 he received Nobel Prize in Chemistry. Since then, PCR is extensively used in molecular detection as well in studying the phylogeny of plant pathogens (Henson and French 1993; Caruso et al. 2003; Pandey et al. 2015; Fang and Ramasamy 2015; Balodi et al. 2017). Different variants of PCR, viz. co-operational PCR, multiplex PCR, multiplex nested RT-PCR and real-time PCR, are proficient in rapid and accurate plant disease diagnosis (Pandey et al. 2015; Yang and Juzwik 2017). Most wilt diseases of pulse crops are caused by *Fusarium* spp. in which conventional approaches of identification are time-consuming and require eminent competence in *Fusarium* taxonomy and physiology (Leslie and Summerell 2006; Thokala et al. 2015). Apart from detection, phenotypic and genotypic characterization of pathogen variants prevalent in particular area is also of great significance in plant disease management. In 2015, Chitten et al. identified *Fusarium* spp. associated with root rot of field peas in North Dakota through PCR using translation elongation factor alpha 1 (TEF-1 α) region. Jimenez-Gasco and Jimenez-Diaz (2003) developed PCR-based detection assay for *Fusarium oxysporum* f. sp. *ciceris*, chickpea wilt pathogen to selectively differentiate pathogenic and nonpathogenic *F. oxysporum* isolates as well as other species and formae speciales of *Fusarium* and *F. oxysporum*, respectively, and each of the *F. oxysporum* f. sp. *ciceris* pathogenic races 0, 1A, 5 and 6. Apart from detection PCR has been employed for studying the diversity of pathogens affecting pulse crops. Dubey et al. (2012) studied the diversity present in *Rhizoctonia solani* infecting different pulse crops in different Indian agroecological regions at molecular level using 23 inter-simple sequence repeats (ISSR) markers, 12 universal rice primers (URPs) and 22 random amplified polymorphic DNA (RAPD).

6.5.2 Real-Time PCR (RT-PCR)

Among PCR techniques, real-time PCR (RT-PCR) has been proven as one of the reliable, sensitive and easy to perform techniques for detection and quantification of soilborne pathogens. This technique allows the real-time monitoring of PCR reaction. RT-PCR-based quantification of soilborne pathogens can provide more accurate and authentic estimation of inoculum load in soil unlike culturing methods which are comparatively less reliable and inaccurate (Mirmajlessi et al. 2015). One of the significant applications of RT-PCR in plant disease diagnostics is simultaneous detection of more than one pathogen when lots of samples are involved (Cooke et al. 2007). Vandemark and Grunwald (2005) applied RT-PCR to establish the relationship between disease severity of Pea root rot and *Aphanomyces euteiches* DNA in soil. Gangneux et al. (2014) developed a rapid and sensitive assay for reliable detection and quantification of *Aphanomyces* in soil. RT-qPCR assay was developed for rapid detection and quantification of *Fusarium* wilt pathogen of *Phaseolus vulgaris*, *F. oxysporum* f. sp. *phaseoli* (Sousa et al. 2014).

6.5.3 Loop-Mediated Isothermal Amplification Assay

Loop-mediated isothermal amplification (LAMP) is an one step amplification assay with great sensitivity and specificity which takes less than an hour to make multiple copies of DNA/RNA (up to 10^9) from very few copies of template under isothermal conditions (Notomi et al. 2000). Four different primers, viz. Forward Inner Primer (FIP), Forward Outer Primer (FOP), Backward Inner Primer (BIP) and Backward Outer Primer (BOP), targeting six distinct region of target gene are used in a LAMP reaction. The overview of different stages of LAMP is available at http://loopamp.eiken.co.jp/e/lamp/anim.html. The technique has been used as a rapid and accurate method for plant disease detection and diagnosis (Tomlinson and Boonham 2008; Khan et al. 2018; Huang et al. 2017). Ghosh et al. (2017) used this novel technique to develop a rapid and sensitive diagnosis for dry root rot of chickpea caused by *R. bataticola* (Taub.) Butler targeted the 5.8S rDNA region of fungus. Rapid diagnosis for *Ascochyta* blight of chickpea pathogen, *Ascochyta rabiei* L. (*A. rabiei*), was also developed through LAMP method with 6.01×10^{-6} ng/μl detection limit based on internal transcribed spacer (ITS) region (Chen et al. 2016). Rapid detection of *F. oxysporum* f. sp. *ciceris* (Foc), chickpea wilt pathogen through LAMP combined with hydroxynaphthol blue (HNB) was performed which developed sky blue colour with Foc DNA but not with negative control (without DNA) or with other fungal DNA (*F. acuminatum*, *F. udum*, *F. solani*, *R. bataticola*, *Alternaria alternata* and *Phytophthora cajani*) (Ghosh et al. 2015).

6.6 Management of Root Rots and Wilt Diseases of Pulse Crops

For disease development by any biotic factor, successful interaction between susceptible host, virulent pathogen and favourable environment is required which remains for a sufficient period of time. The interference and manipulation of any of these components during disease development before the occurrence of sufficient loss to reduce the disease level below economic injury level with minimum harm to the environment is the basic principle of plant disease management (Katan 2017). This cannot be achieved by just adopting a single tactics but to amalgamate all the approaches, viz. cultural and mechanical methods, chemical methods, biological control, using host plant resistance, etc. in the framework of "integrated plant disease management (IDM)". IDM can be defined as a "decision-based process involving coordinated use of multiple tactics for optimizing the control of all classes of pests (insects, pathogens, weeds, vertebrates) in an ecologically and economically sound manner" (Prokopy 2003). As soilborne pathogens produce their resting structures in soil, therefore they are influenced by biotic and abiotic factors of soil which changes with agricultural practices that are applied to the soil. In case of soilborne pathogens, it is not always necessary that soil inoculum is the only and major source of inoculum and thus makes the management even more tedious. There are four foremost steps in soilborne disease management, viz., prevention of introduction and establishment of pathogen to newer cultivating areas, reduction of pathogen population below economic injury level, improvement of natural suppressiveness of soil and least manipulation of natural biological and physical properties of the soil (Chellemi et al. 2016).

6.6.1 *Cultural and Mechanical Methods*

By adopting good cultural practices, one can maintain an environment favourable for crop but not for the disease development. The present cultural practices or traditions which are followed today to control soilborne pathogens are the result of numerous observations and long-term experience generated through trials and errors though due to the availability of effective chemical, the interest in cultural practices is lost among the growers (Katan 2010). Howbeit, with increasing concern of deteriorating environment and popularization of IDM concept, the interest in cultural practices has been again emerged. Application of diverse cultural practices, viz. intercropping/mixed cropping, crop rotation, field sanitation, adjusting sowing times, etc., are advocated as effective tools for soilborne disease management (Juroszek and von Tiedemann 2011; Pandey et al. 2018). However the survival period of wilt and root rot pathogens in soil is very long, therefore at least 5 or 7 years of rotations are required to prevent the building up of pathogen population to a level causing damage above economic injury level. Intercropping of pigeon pea with sorghum at 1:1 was found effective in managing *Fusarium* wilt when integrated with other management approaches (Prasad et al. 2012). The sorghum's root exudates which include hydrocyanic acid and tannins are reported to affect the mycelial growth and conidial germination of *Fusarium* spp. in soil (Rangaswami and Balasubramanian 1963; Odunfa 1979). High temperature during maturity of chickpea can be prevented by timely or early sowing of chickpea; moreover, when supplemented with timely irrigation, DRR incidence is further reduced (Sharma et al. 2015). Yaqu and Shahzad (2009) has observed less disease incidence due to the use of plastic mulching which led to sclerotial mortality of *M. phaseolina*, the dry root rot pathogen. In lentil, sowing in the first week of December at 2 cm depth results in least wilt severity and highest grain yield (Sallam and Monaim 2012).

6.6.2 Chemical Control

The soilborne nature of both wilt and root rot causing pathogens, development of resistance to chemicals in pathogenic isolates over the time and zero possibility of treating soil at a large scale make chemical management less worthy than cultural practices. But still there are reports where chemicals are in use. Generally, foliar sprays are found to be less effective in management of soilborne pathogens as compared to seed and soil treatment. The management of soilborne pathogens starts with the chemical treatment of soil and commonly used chemicals are metalaxyl, diazoben, pentachloronitrobenzene, captan and chloroneb (Veena et al. 2014). The foliar spray of Fosetyl-aluminium has been reported to control soilborne pathogens. After soil treatment, the seed treatment with various fungicides alone or in combination is also in use; seed treatment with tebuconazole at 1 ml/kg (for gram wilt), difenoconazole, carbendazim, thiram, mixture of benomyl and thiram and a combination of carbendazim + thiophanate (0.15 + 0.10%), carbendazim 12% + mancozeb 63% WP has been recommended by various workers for the control of root rots and wilts (Sinha et al. 2018; Golakiya et al. 2018; Durga et al. 2014). Seed treatment with thiram + PCNB or thiram + carboxin was reported to keep check over lentil wilt, while for the management of pea wilt tebuconazole/metalaxyl M + difenoconazole/imidacloprid + tebuconazole were recommended. For the control of DRR of chickpea, seed treatment with carbendazim and thiophanate methyl was found to be effective (Sharma and Kumara 2017). Overall, the management of both wilts and root rot pathogens are the same.

6.6.3 Biological Control

Keeping in view the environmental losses, reduction in beneficial soil microflora and microfauna, residual effects due to excessive use of pesticides and development of resistance in pathogens, the biological control offers an attractive and ecofriendly alternative approach for plant disease management (Chandrashekara et al. 2012; Singh 2014). Biological control has been defined as "the action of parasites, predators, or pathogens in maintaining another organism's population density at a lower average than would occur in their absence" (DeBach 1964). Several potential biocontrol agents (BCAs) have been identified for the management of soilborne diseases. The potential BCAs identified are *Gliocladium*, *Trichoderma*, nonpathogenic *Fusarium*, *Bacillus*, fluorescent *Pseudomonas*, *Streptomyces*, etc. (Harman and Kubhicek 1998; Benhamou et al. 2012; Bochow et al. 1997; Weller 1988). King and Parke (1993) applied *Pseudomonas cepacia* strain AMMD as seed treatment and achieved control in case of four pea cultivars against *Pythium* sp. causing damping off and *Aphanomyces* root rot. Root rot disease in chickpea due to *Meloidogyne incognita* and *Macrophomina phaseolina* was least when all the three phosphate solubilizing bacteria, viz. *Pseudomonas aeruginosa* (isolate Pa28), *Aspergillus awamori*, and *Glomus intraradices* were inoculated together greatest increase in the plant growth (Siddiqui and Akhtar 2007). Shahid and Khan (2016) evaluated the biocontrol efficiency of different fungi and bacteria, viz. *Trichoderma harzianum*, *T. reesei*, *Aspergillus niger* and *Bacillus subtilis* against *M. phaseolina*, DRR of mungbean pathogen, and found *T. harzianum* and *B. subtilis* as best BCA in managing the disease as well as improving the plant growth and yield of mungbean. *Actinobacteria Streptomyces* has also been identified as potential BCA against *Aphanomyces euteiches*, the causal agent of pea root rot based on in vitro antimicrobial activity assay followed by identification based on 16S rDNA analysis and morphological and chemical characteristics (Oubaha et al. 2018). For the management of wilt and root rot diseases of pulse crops, seed treatment with *T. viride* at 4 g/kg or *P. fluorescens* at 10 g/kg of seed or spot drenching with *P. fluorescens* / *T. viride* 2.5 kg/ha with 50 kg Farm Yard Manure (FYM) has been recommended.

6.6.4 Host Plant Resistance

Host resistance offers the most economic and environment-friendly method of plant disease management. In case of soilborne diseases, use of resistant varieties is the most practical approach for their management. Serious efforts are being taken in the direction of finding new sources of resistance in wild relatives of cultivated pulse crops, mapping of resistance genes/quantitative trait loci (QTL) and identifying genetic markers linked with identified resistant (R)genes/QTLs for application of marker-assisted selection (MAS) in resistance breeding programmes. The application of MAS by identifying molecular markers linked to R genes against different pathogen races can accelerate the resistance breeding programme (Winter and Kahl 1995). For resistance breeding programmes, a clear picture of the existing pathogenic variability and races present in the target area is the prerequisite. A lot of conventional as well as molecular breeding programmes have been conducted worldwide in developing resistant chickpea cultivars. The existence of race 1, 2, 3 and 4 was confirmed by Haware and Nene (1982) in India using ten chickpea differential lines.

Jimenez-Diaz et al. (1993) studied the pathogenic variability of 107 *F. oxysporum* f. sp. *ciceris* (Foc) isolates from Algeria, California, Morocco, Tunisia, Spain and Italy and screened 2702 kabuli lines procured from ICARDA for resistance against *Fusarium* wilt. Different workers have mapped resistant genes for Foc race 1, 2, 4 and 5 on the same linkage group (Simon and Muehlbauer 1996; Ratnaparkhe et al. 1998). Benko-Iseppon et al. (2003) identified molecular markers closely linked to *Fusarium* R genes in chickpea through bulked segregant analysis (BSA) which showed significant alignments to pathogenesis-related (PR) genes located on 1 and 5 chromosomes of *Arabidopsis*. Iftikhar and Ilyas (2000) found only ICCV 97112 was found resistant out of 108 chickpea germplasms screened for resistance against DRR. Gangwar et al. (2002), Prajapati et al. (2003), Pande et al. (2006) and Khan et al. (2013) reported few resistance sources in chickpea against DRR. Marker-assisted backcrossing programmes were undertaken to introgress resistance against *Ascochyta* blight and *Fusarium* wilt in chickpea cultivar, C 214 targeting two QTL regions, viz. ABQTL-I and ABQTL-II and foc1 locus. Foreground selection for foc1 locus in case of *Fusarium* wilt Race 1 was conducted using six markers, viz. TA194, TR19, GA16, TAA60, TS82 and TA110, while in case of *Ascochyta* blight, eight markers, viz. GAA47, TA2, TA194, TR58, TS82, TA130, SCY17 and GA16, linked to ABQTL-I and ABQTL-II were used (Varshney et al. 2014).

In mungbean breeding programmes against disease resistance, MAS has not been much exploited; however molecular markers against major resistant (R) genes or QTLs against fungal diseases like powdery mildew and *Cercospora* leaf spot have been identified, but no associated molecular markers or R gene or QTLs were reported for DRR of mungbean (Pandey et al. 2018). In case of *Fusarium* wilt of pea, four races, viz. 1, 2, 5 and 6, of *F. oxysporum* f.sp. *pisi* (Fop) were recognized by Kraft and Pfleger (2001). *Fusarium* wilt resistance against majority of Fop races is governed by single gene (Coyne et al. 2000; Grajal-Martin and Muehlbauer 2002; McClendon et al. 2002; Kwon et al. 2013). However, resistance against Fop race 2 is quantitative (Bani et al. 2011; McPhee et al. 2012). Single gene *Fw* was located on linkage group III which confers resistance against Fop race 1 in pea (Kwon et al. 2013). Kwon et al. (2013) identified three tightly linked markers to *Fw* locus, viz. Fw_Trap_480, Fw_Trap_340 and Fw_Trap_220, which were only 1.2 cM away from the locus. These markers were found to be suitable for their use in MAS for Fop race 1 breeding programmes. A genetic linkage map was constructed for *Fusarium* wilt resistance and localized on linkage group 6 in lentil based on micro-satellite markers mapping identified from genomic library of lentil (*Lens culinaris* Medis.) (Hamwieh et al. 2005).

6.7 Conclusion

The Indian agriculture is still struggling due to incidence of pests and diseases resulting in huge crop losses. Though fungal pathogens are known to incite plant diseases since 1807, still phytopathogenic fungi take a heavy toll on crop production worldwide. Out of different classes of fungal pathogens, soilborne pathogens manage to remain as the most notorious one due to various factors like challenges that are there in timely detection and therefore management, long survival period in soil, complex nature of diseases caused by them due to involvement of multiple microorganisms and nematodes as well. Root rots and wilts are the major limiting factors of pulse crop production in India. Despite the considerable application of chemicals and other management approaches that include cultural, biological and exploitation of host resistance, these diseases continue to be a constraint in pulse crop production. The nature of these pathogens, various climatic factors affecting the incidence and disease development caused by these pathogens are already known; however extensive studies are required to elucidate the infection process and determine the pathogenic and genetic variation, spatial and temporal distribution of causal pathogens and resistance mechanism in host plants. Moreover, the application of advanced molecular tools in timely and precise detection and diagnosis of root rots and wilt pathogens is very limited. For sustainable management of root rot and wilt in pulse crops, reliable marker-assisted resistance breeding programmes suitable for broader geographical areas using tightly R-gene linked markers are required. Different omics approaches must be employed to identify the molecular mechanism of resistance and key molecular factors playing role in governing resistance against root rots and wilts in already identified resistant lines.

References

Agarwal DK, Billore SD, Sharma AN, Dupare BU, Srivastava SK. Soybean: introduction, improvement, and utilization in India—problems and prospects. Agric Res. 2013;2(4):293–300.

Ahsan S, Kumar M, Upadhyay JP, Hussain A, Gupta PK, Singh A. Effect of different doses of *Trichoderma harzianum* and fungicides for the management of collar rot of chickpea caused by *Sclerotium rolfsii*. Int J Pure App Biosci. 2018;6(1):1656–60. https://doi.org/10.18782/2320-7051.5926.

Arunodhayam K, Reddy NPE, Madhuri V. Pathogenicity and management of *Fusarium* wilt of chickpea, *Cicer arietinum* L.–A review. Current Biotica. 2014;7(4):343–58.

Arya A, Kushwaha KPS. Evaluation of chemicals for the management of lentil wilt, caused by *Fusarium oxysporum* f.sp. *lentis*. J Pharmacogn Phytochemistry. 2018;7(5):2320–3.

Back MA, Haydock PPJ, Jenkinson P. Disease complexes involving plant parasitic nematodes and soil-borne pathogens. Plant Pathol. 2002;51:683–97.

Balodi R, Bisht S, Ghatak A, Rao KH. Plant disease diagnosis: technological advancements and challenges. Indian Phytopath. 2017;70:275–81.

Bani M, Rubiales D, Rispail N. A detailed evaluation method to identify sources of quantitative resistance to *Fusarium oxysporum* f. sp. *pisi* race 2 within a *Pisum* spp. germplasm collection. Plant Pathol. 2011;61:532–42.

Benhamou N, le Floch G, Vallance J, Gerbore J, Grizard D, Rey P. *Pythium oligandrum*: an example of opportunistic success. Microbiol. 2012;158:2679–94. https://doi.org/10.1099/mic.0.061457-0. Epub 2012 Sep 13.

Benko-Iseppon A-M, Winter P, Huettel B, Staginnus C, Muehlbauer FJ, Kahl G. Molecular markers closely linked to *Fusarium* resistance genes in chickpea show significant alignments to pathogenesis-related genes located on Arabidopsis chromosomes 1 and 5. Theor Appl Genet. 2003;107:379–86.

Bhatt J, Vadhera I. Histopathological studies on cohabitation of *Pratylenchus thornei* and *Rhizoctonia bataticola* on chickpea (*Cicer arietinum* L.). Adv Plant Sci. 1997;10:33–7.

Bochow H, El-Sayed SF, Junge H, Stavropoulou A, Schmiedeknecht G. Use of *Bacillus subtilis* as biocontrol agent for Salt-stress tolerance induction by *Bacillus subtilis* FZB24 seed treatment in tropical vegetable field crops, and its mode of action. J Plant Dis Prot. 1997;108:21–30.

Butler EJ. The wilt disease of Pigeonpea and Pepper. Agriculture Journal of India. 1906;1: 25–26.

Butler EJ. Fungi and Diseases in Plants. Thacker Spink and Co.; 1918.

Caruso P, Bertolini E, Cambra M, Lopez MM. A new and co-operational polymerase chain reaction (Co-PCR) for rapid detection of *Ralstonia solanacearum* in water. J Microbiol Methods. 2003;55:257–72.

Chand H, Khirbat SK. Chickpea wilt and its management - a review. Agric Rev. 2009;30(1):1–12.

Chandrashekara K, Chandrashekara C, Chakravathi M, Manivannan S. Biological control of plant disease. In: Singh VK, Singh Y, Singh A, editors. Eco-friendly innovative approaches in plant disease management: International Book Distributors; Uttarakhand: 2012.

Chaudhary B. Studies on wilt of Pigeonpea caused by *Fusarium udum butler*. M.Sc. Thesis. Department of Plant Pathology, College of Agriculture, Jabalpur 482004 Jawaharlal Nehru KrishiVishwaVidyalaya, Jabalpur, Madhya Pradesh; 2016. pp20.

Chaudhary RG, Dhar V, Singh RK. Association of fungi with wilt complex of lentil at different crop growth stages and moisture regimes. Arch Phytopath Plant Prot. 2009;42:340–3. https://doi.org/10.1080/03235400601070397.

Chellemi DO, Gamliel A, Katan J, Subbarao KV. Development and deployment of system-based approaches for the management of soil borne plant pathogens. Phytopathology. 2016;106:216–25.

Chen X, Ma L, Qiang S, Maa D. Development of a loop-mediated isothermal amplification method for the rapid diagnosis of *Ascochyta rabiei* L. in chickpeas. Sci Rep. 2016;6:1–10.

Chennakesavulu M, Reddikumar M, Reddy NPE. Evaluation of different fungicides and their compatibility with *Pseudomonas fluorescens* in the control of red gram wilt incited by *Fusarium udum*. JBC. 2013;27:354–61.

Cleary M, Sturrock R, Hodge J. Laminated root disease- stand establishment decision aid. JEM. 2011;12(2):17–20.

Cooke DEL, Schena L, Cacciola SO. Tools to detect identify and monitor *Phytophthora* species in natural ecosystems. J Plant Pathol. 2007;89:13–28.

Coyne CJ, Inglis DA, Whitehead SJ, Muehlbauer FJ. Chromosomal location of Fwf, the Fusarium wilt race 5 resistance gene in *pisum sativum*. Pisum Genet. 2000;32:20–2.

Cui BK, Dai YC, He SH, Zhou LW, et al. A Novel *Phellinidium* sp. causes laminated root rot on qilian juniper (*Sabina przewalskii*) in Northwest China. Plant Dis. 2014;99:39–43.

Dastur JF. Gram wilts in central provinces. Agric Liv Stk. 1935; 5:615–628.

DeBach P. Biological control of insect pests and weeds. New York: Reihold; 1964.

DeShields JB, Bomberger RA, Woodhall JW, Wheeler DL, Moroz N, Johnson DA, et al. On-site molecular detection of soil-borne phytopathogens using a portable real-time PCR system. J Vis Exp. 2018;(132):1–11.

Dhingra OD, Sinclair JB. An annotated bibliography of *Macrophomina phaseolina*. 1905–1975. Minas Gerais, Brazil: Universidade Federal de Vicosa; 1977.

Diourte M, Starr JL, Jeger MJ, Stack JP, Rosenow DT. Charcoal rot (*Macrophomina phaseolina*) resistance and the effects of water stress on disease development in sorghum. Plant Pathol. 1995;44:196–202.

Dubey SC, Tripathi A, Upadhyay BK. Molecular diversity analysis of *Rhizoctonia solani* isolates infecting various pulse crops in different agro-ecological regions of India. Folia Microbiol (Praha). 2012;57:513–24.

Durga KK, Bharathi V, Rani MS, Reddy AV. Chemical and biological management of wilt and root rot of chickpea. 2nd International Conference on Agricultural & Horticultural Sciences. 2014; 2:170. https://doi.org/10.4172/2168-9881.S1.007.

El Karkouri A, El Hassani FZ, El Mzibri M, et al. Isolation and identification of an actinomycete strain with a biocontrol effect on the phytopathogenic *Erwinia chrysanthemi* 3937VIII responsible for soft rot disease. Ann Microbiol. 2010;60(2):263–8.

Fang Y, Ramasamy RP. Current and prospective methods for plant disease detection. Biosensors. 2015;4:537–61.

FAOSTAT. Agriculture; 2017. Available from: http://faostat.fao.org

Gangneux C, Cannesan MA, Bressan M, Castel L, Moussart A, Vicré-Gibouin M, et al. A sensitive assay for rapid detection and quantification of *Aphanomyces euteiches* in soil. Phytopathology. 2014;104:1138–47. https://doi.org/10.1094/PHYTO-09-13-0265-R.

Gangwar RK, Prajapati RK, Srivastava SSL, Kumar K. Resistance in chickpea germplasm against the dry root rot. Ann Plant Protect Sci. 2002;10:393–4.

Garkoti A, Kumar S, Lal M, Singh V. Major diseases of lentil: epidemiology and disease management-a review. Agri. 2013;1(1):62–4.

Ghosh R, Sharma M, Telangre R, Pande S. Occurrence and distribution of chickpea diseases in central and southern parts of India. Am J Plant Sci. 2013;4:940–4. https://doi.org/10.4236/ajps.2013.44116.

Ghosh R, Nagavardhini A, Sengupta A, Sharma M. Development of loop-mediated isothermal amplification (LAMP) assay for rapid detection of *Fusarium oxysporum* f. sp. *ciceris* – wilt pathogen of chickpea. BMC Res Notes. 2015;8:1–10. https://doi.org/10.1186/s13104-015-0997-z.

Ghosh R, Tarafdar A, Sharma M. Rapid and sensitive diagnoses of dry root rot pathogen of chickpea (*Rhizoctonia bataticola* (Taub.) Butler) using loop-mediated isothermal amplification assay. Sci Rep. 2017;7:1–12. https://doi.org/10.1038/srep42737.

Golakiya BB, Bhimani MD, Akbari LF. Efficacy of different fungicides for the management of chickpea wilt (*Fusarium oxysporum* f. sp. *ciceri*). Int J Chem Stud. 2018;6:199–205.

Grajal-Martin MJ, Muehlbauer FJ. Genomic location of the Fw gene for resistance to *Fusarium* wilt race 1 in peas. J Hered. 2002;93:291–3.

Hamwieh A, Udupa SM, Choumane W, Sarker A, Dreyer F, Jung C, et al. A genetic linkage map of Lens sp. based on microsatellite and AFLP markers and the localization of *Fusarium* vascular wilt resistance. Theor Appl Genet. 2005;110:669–77.

Harman GE, Kubicek CP. Trichoderma and Gliocladium, Vol. 2. Enzymes, Biological Control and Commercial Applications. Taylor and Francis; 1998.

Hasan R, Khan DN. Diagnosis of pulse production and consumption in Uttar Pradesh: an inter regional analysis. Int J Curr Microbiol App Sci. 2018;7(7):860–5.

Haware MP, Nene YL. Races of *Fusarium oxysporum* f. sp. *ciceri*. Plant Dis. 1982;66:809–810.

Heffer Link V, Powelson ML, Johnson KB. Oomycetes: The Plant Health Instructor. 2002; https://doi.org/10.1094/PHI-I-2002-0225-01.

Henson J, French R. The polymerase chain reaction and plant disease diagnosis. Annu Rev Phytopathol. 1993;31:81–109.

Holliday P, Punithalingam E. *Macrophomina phaseolina* CMI description of pathogenic fungi and bacteria. No. 275. Farnham Royal: Commonwealth Agricultural Bureaux; 1970.

Huang W, Zhang H, Xu J, Wang S, Kong X, Ding W, Xu J, Feng J. Loop-mediated isothermal amplification method for the rapid detection of *Ralstonia solanacearum* phylotype I mulberry strains in China. Front Plant Sci. 2017;8:1–10. https://doi.org/10.3389/fpls.2017.00076.

Iftikhar K, Ilyas MB. Screening of chickpea germplasm against dry root rot disease (*Macrophomina phaseolina*) in pots/glass house. Pakistan J Phytopathol. 2000;12:66–70.

Jendoubi W, Bouhadida M, Boukteb A, Beji M, Kharrat M. *Fusarium* wilt affecting chickpea crop. Agriculture. 2017;7:1–16. https://doi.org/10.3390/agriculture7030023.

Jimenez-Díaz RM, Castillo P, Jimenez-Gasco MM, Landa BB, Navas-Cortes JA. Fusarium wilt of chickpeas: biology, ecology and management. Crop Prot. 2015;73:16–27.

Jimenez-Gasco MM, Jimenez-Diaz RM. Development of a specific polymerase chain reaction-based assay for the identification of *Fusarium oxysporum* f. sp. *ciceris* and its pathogenic races 0, 1A, 5 and 6. Phytopathology. 2003;93:200–9.

Jimenez-Diaz RM, Alcala-Jimenez AR, Hervas A, Trapero-Casas JL. Pathogenic variability and host resistance in the *Fusarium oxysporum* f. sp. *ciceris/Cicer arietinum* pathosystem. In: Arseniuk E, Goral T, editors. Fusarium Mycotoxins, Taxonomy, Pathogenicity and Host Resistance. Proceedings of the 3rd European Seminar. Radzikov, Poland, Plant Breeding and Acclimatization Institute; 1993. p 87–94.

Jones FR, Linford MB. Pea Disease Survey in Wisconsin. Wisc Agric Exp Sta Res Bull; 1925.

Juroszek P, von Tiedemann A. Potential strategies and future requirements for plant disease management under a changing climate. Plant Pathol. 2011;60:100–12. https://doi.org/10.1111/j.1365-3059.2010.02410.x.

Kadam AM, Chavan SS, Dhutraj DN, Rewale KA. Survey of dry root rot of chickpea incidence in Marathwada region. J Pharmacogn Phytochem. 2018;SP1:3004–8.

Katan J. Cultural approaches for disease management: present status and future prospects. J Plant Pathol. 2010;92:S4.7–9.

Katan J. Diseases caused by soilborne pathogens: biology, management and challenges. J Plant Pathol. 2017;99:305–15.

Katoch S, Rana SK, Sharma PN. Application of PCR based diagnostics in the exploration of *Parastagonospora nodorum* prevalence in wheat growing regions of Himachal Pradesh. J Plant Biochem Biotechnol. 2019;28(2):169–75. https://doi.org/10.1007/s13562-018-0481-7.

Khan RA, Bhat TA, Kumar K. Screening of chickpea (*Cicer arietinum* L.) germplasm lines against dry root rot caused by *Rhizoctonia bataticola* (taub.) Butler. Asian J Pharm Clin Res. 2013;6:211–2.

Khan M, Wang R, Li B, Liu P, Weng Q, Chen Q. Comparative evaluation of the LAMP assay and PCR-based assays for the rapid detection of *Alternaria solani*. Front Microbiol. 2018;9:1–11. https://doi.org/10.3389/fmicb.2018.02089.

Kheni J, Kothari V, Rathod K, Rathod V, Padhiyar S, Bhalara R, et al. Morphological and metabolic characterization of wilt disease (*Fusarium oxysporum* f. sp. *ciceri*) in chickpea (*Cicer arietinum* L.). Res J Agric Biol Sci. 2017;5:12–9.

King EB, Parke JL. Biocontrol of Aphanomyces root rot and Pythium damping-off by *Pseudomonas cepacia* AMMD on four Pea cultivars. Plant Dis. 1993;77:1185–8.

Kraft JM, Pfleger FL. Compendium of pea diseases and pests. 2nd ed. St. Paul: The American Phytopathological Society; 2001.

Kumari S, Khanna V. Biological management of vascular wilt of chickpea (*Cicer arietinum* L.) incited by *Fusarium oxysporum f. sp. ciceris* by antagonistic rhizobacteria co-inoculated with native *Mesorhizobium*. Int J Currt Microbiol Appl Sci. 2018;7(1):920–41.

Kushi KK, Khare MN. Comparative efficacy of five methods to detect *Macrophomina phaseolina* with sesamum seeds. Indian Phytopath. 1978;31:258–9.

Kushwaha SK. Studies on collar rot of lentil caused by *Sclerotium rolfsii* Sacc. M.Sc. Thesis, Jawaharlal Nehru Krishi Vishwa Vidyalaya, Jabalpur; 2016. pp. 1–3.

Kwon SJ, Smykal P, Hu J, Wang M, Kim S-J, McGee RJ, et al. User-friendly markers linked to *Fusarium* wilt race 1 resistance Fw gene for marker-assisted selection in pea. Plant Breed. 2013;132:642–8.

Legg JP, Jeremiah SC, Obiero HM, Maruthi MN, Ndyetabula I, Okao-Okuja G, et al. Comparing the regional epidemiology of the cassava mosaic and cassava brown streak virus pandemics in Africa. Virus Res. 2011;159:161–70.

Leslie JF, Summerell BA. The *Fusarium* laboratory manual. Ames, IA: Blackwell Publishing; 2006.

Linford MB. A *Fusarium* Wilt of Peas in Wisconsin. Wisc Agric Exp Sta Res Bull; 1928.

Lodhi NAK, Abbas A, Waris W, Asad M, Aslam MM. Chickpea wilt and its management strategies – a review paper. Imp J Interdiscip Res. 2006;2(11):1281–90.

Mallaiah B, Rao VK. Integrated management of dry root rot of green gram (*Vigna radiate* (L.) Wilczek) incited by *Macrophomina phaseolina* (Tassi.) Goid. Int J Trop Agri. 2016;34(3):607–14.

Maruti SAS, Sunkad G, Mahalinga D, Patil MG. Incidence of dry root rot of pigeon pea in North Eastern Karnataka, India. Int J Curr Microbiol App Sci. 2017;6(11):1071–8. https://doi.org/10.20546/ijcmas.2017.611.126.

McClendon T, Inglis A, McPhee E, Coyne J. DNA markers linked to *Fusarium* wilt race 1 resistance in pea. J Am Soc Sci. 2002;127:602–7.
McPhee KE, Inglis DA, Gundersen B, Coyne CJ. Mapping QTL for *Fusarium* wilt race 2 partial resistance in pea (*Pisum sativum*). Plant Breed. 2012;131:300–6.
Meena R. Studies on integrated management of pigeonpea wilt caused by *Fusarium oxysporum* f. sp. *Udum.M.Sc.* Thesis. Department of Plant Pathology, Rajmata Vijayaraje Scindia Krishi Vishwa Vidyalaya B.M. College of Agriculture, Khandwa (M.P); 2016, p. 15.
Meyer WA. Factors Affecting Charcoal Rot of Soybean Seedlings. Phytopathology 1974;64(6):845
Mirmajlessi SM, Loit E, Mand M, Mansouripour SM. Real-time PCR applied to study on plant pathogens: potential applications in diagnosis – a review. Plant Protect Sci. 2015;51:177–90.
Narayan P, Kumar S. Constraints of growth in area production and productivity of pulses in India: an analytical approach to major pulses. Indian J Agric Res. 2015;49:114–24.
Nath N, Ahmed AU, Aminuzzaman FM. Morphological and physiological variation of *Fusarium oxysporum* f. sp. *ciceri* isolates causing wilt disease in chickpea. IJEAB. 2017;2:202–12. https://doi.org/10.22161/ijeab/2.1.25.
Nongmaithem N, Basudha CH, Sharma SK. Incidence of rust, powdery mildew and wilt in pea and broad bean plant of Manipur, India. Int J Curr Microbiol App Sci. 2017;6:2611–6.
Notomi T, Okayama H, Masubuchi H, Yonekawa T, Watanabe K, Amino N, Hase T. Loop-mediated isothermal amplification of DNA. Nucleic Acids Res. 2000;28:E63.
Odunfa VSA. Free amino acids in the seed and root exudates in relation to the nitrogen requirements of rhizosphere soil fusaria. Plant Soil. 1979;52:491–9.
Oubaha B, Nafis A, Baz M, Mauch F, Barakate M. The potential of antagonistic Moroccan *Streptomyces* isolates for the biological control of damping-off disease of pea (*Pisum sativum* L.) caused by *Aphanomyces euteiches*. J Phytopathol. 2018;167(2):1–9. https://doi.org/10.1111/jph.12775.
Pande S, Kishore GK, Upadhyaya HD, Rao JN. Identification of sources of multiple disease resistance in mini-core collection of chickpea. Plant Dis. 2006;90:1214–8.
Pande S, Sharma M, Guvvala G. An updated review of biology, pathogenicity, epidemiology and management of wilt disease of pigeonpea (*Cajanus cajan* (L.) Millsp.). J Food Legumes. 2013;26:1–14.
Pandey P, Pandey NS, Shamim M, Srivastava D, Dwivedi DK, Awasthi LP, Singh KN. Molecular tools and techniques for detection and diagnosis of plant pathogens. In: Awasthi LP, editor. Recent advances in the diagnosis and management of plant diseases. New Delhi: Springer; 2015.
Pandey RN, Gohel NM, Jaisani P. Management of wilt and root rot of chickpea caused by *Fusarium oxysporum* f. sp. *Ciceri* and *Macrophomina phaseolina* through seed biopriming and soil application of bio-agents. Int J Curr Microbiol App Sci. 2017;6:2516–22. https://doi.org/10.20546/ijcmas.2017.605.282.
Pandey AK, Burlakoti RR, Kenyon L, Nair RM. Perspectives and challenges for sustainable management of fungal diseases of mungbean [*Vigna radiata* (L.) R. Wilczek var. radiata]: a review. Front Environ Sci. 2018;6:53. https://doi.org/10.3389/fenvs.2018.00053.
Patil M, Gupta O, Rathod PK. Morphological, cultural and pathogenic variation in races and variant of *F. Oxysporum* f. sp. *ciceri* from seven locations of central zone of India. IJAPSA. 2017:66–74.
Patra S, Biswas MK. Studies on cultural, morphological and pathogenic variability among the isolates of *Fusarium oxysporum* f. Sp. *Ciceri* causing wilt of chickpea. Int J Pl An Environ Sci. 2016;7:11–6.
Prajapati RK, Gangwar RK, Srivastava SSL. Resistant sources of chickpea against dry root rot. Farm Sci J. 2003;12:86.
Prasad PS, Saifulla M, Mahesh M, Kumar GNV. Management of pigeonpea wilt caused by *Fusarium udum* Butler through integrated approaches. Biol Control. 2012;26:361–7.
Prokopy RJ. Two decades of bottom-up, ecologically based pest management in a small commercial apple orchard in Massachusetts. Agric Ecosyst Environ. 2003;94:299–309.

Rai B, Upadhyay RS. Gibberella Indica: The Perfect State of *Fusarium udum*. Mycologia. 1982; 74: 343–346.

Ram RM, Singh HB. *Rhizoctonia bataticola*: a serious threat to chickpea production. Int J Chem Stud. 2018;6:715–23.

Ramezani M, Shier WT, Abbes HK, Tones JL, Baird RE, Sciumbato GL. Soybean charcoal rot disease fungus *Macrophomina phaseolina* in Mississippi produces the phytotoxin botryodiplodin but no detectable phaseolinone. J Nat Prod. 2007;70:128–9.

Rangaswami G, Balasubramanian A. Release of hydrocyanic acid by sorghum roots and its influence on the rhizosphere microfiora and plant pathogenic fungi. Indian J Expt Biol. 1963;1:215.

Rao JK. Studies on survey of *Fusarium* wilt of pea in Eastern Uttar Pradesh. Int J Life Sci. 2014;2:359–62.

Ratnaparkhe MB, Santra DK, Tullu A, Muehlbauer FJ. Inheritance of inter-simple-sequence-repeat polymorphisms and linkage with a *Fusarium* wilt resistance gene in chickpea. Theor Appl Genet. 1998;96:348–53.

Rawat L, Singh Y, Kumar B, Shulkla A. Management of Rhizoctonia root rot of pea (*Pisum sativum* L.) by integrated biological and chemical approach. Int J Agri Sci. 2014;10:108–14.

Ray M, Dash S, Shahbazi S, Achary KG, Nayak S, Singh S. LWT – food Sci. Technol. 2016;66:546–52.

Sallam NMA, Monaim MFA. Influence of some agricultural practices on suppression of lentil wilt disease. Plant Pathol J. 2012;11:32–7.

Sankar PM, Vanitha S, Kamalakannan A, Raju PA, Jeyakumar P. Prevalence of *Fusarium oxysporum* f. sp. *ciceris* causing wilt in chickpea and its pathogenic, cultural and morphological characterization. Int J Curr Microbiol App Sci. 2018;7(2):1301–13.

Sarkar M, Bhattacharyya PK. Biological control of root rot of green gram caused by *Macrophomina phaseolina* by antagonistic microorganisms. J Mycopathol Res. 2008;46:233–7.

Saxena KB, Kumar RV, Saxena RK, Sharma M, Srivastava RK, Sultana R, et al. Identification of dominant and recessive genes for resistance to Fusarium wilt in pigeonpea and their implication in breeding hybrids. Euphytica. 2012;188:221–7. https://doi.org/10.1007/s10681-012-0700-6.

Shahid S, Khan MR. Biological control of root-rot on mung bean plants incited by *Macrophomina phaseolina* through microbial antagonists. Plant Pathol J. 2016;15:27–39.

Sharma P. Alarming occurrence of *Fusarium* wilt disease in pea (*Pisum sativum* L.) cultivations of Jabalpur district in Central India revealed by an array of pathogenicity tests. Agric Biol J N Am. 2011;2:981–94.

Sharma OP, Kumara M. Management of dry root rot disease [*Rhizoctonia bataticola*] of chickpea through fungicides. Int J Chem Stud. 2017;5:45–7.

Sharma M, Pande S. Unravelling effects of temperature and soil moisture stress response on development of dry root rot [*Rhizoctonia bataticola* (Taub.)] butler in chickpea. Am J Plant Sci. 2013;4:584–9.

Sharma YK, Gaur RB, Bisnoi HR. Cultural, morphological and physiological variability in *Macrophomina phaseolina*. J Mycol Plant Pathol. 2004;34:532–4.

Sharma M, Ghosh R, Suresh P. Dry root rot (*Rhizoctonia bataticola* (Taub.) Butler): an emerging disease of chickpea – where do we stand? Arch Phytopathol Plant Prot. 2015;48:797–812.

Siddiqui ZA, Akhtar MS. Biocontrol of a chickpea root rot disease complex with phosphate- solubilizing microorganisms. J Plant Pathol. 2007;89:67–77. https://doi.org/10.4454/jpp.v89i1.725.

Simon CJ, Muehlbauer FJ. Construction of a chickpea linkage map and its comparison with maps of pea and lentil. J Hered. 1996;88(2):115–9.

Singh HB. Management of plant pathogens with microorganisms. Proc Indian Natn Sci Acad. 2014;80:443–54. https://doi.org/10.16943/ptinsa/2014/v80i2/55120.

Singh NP. Pulses as a candidate crops for doubling farmers' income. Indian Farming. 2018;68:36–43.

Singh PJ, Mehrotra RS. Penetration and invasion of gram roots by *Rhizoctonia bataticola*. Indian Phytopathol. 1982;35:336–8.

Singh SK, Nene YL, Reddy MV. Some histopathological observations of chickpea roots infected by *Rhizoctonia bataticola*. Int Chick News. 1990;23:24–5.

Singh J, Mishra KK, Singh AK. Current status of web blight of mung bean. An Asian J Soil Sci. 2013;8(2):495–504.

Singh AK, Singh SS, Prakash V, Kumar S, Dwivedi SK. Pulses production in India: present status, bottleneck and way forward. J Agri Search. 2015;2:75–83.

Singh D, Sinha B, Rai VP, Singh MN, Singh DK, Kumar R, et al. Genetics of *Fusarium* wilt resistance in pigeonpea (*Cajanus cajan*) and efficacy of associated SSR markers. Plant Pathol J. 2016;32:95–101. https://doi.org/10.5423/PPJ.OA.09.2015.0182.

Singh P, Shahi B, Singh KM. Trends of pulses production, consumption and import in India: current scenario and strategies. MPRA; 2017a. https://mpra.ub.uni-muenchen.de/81589/2017.

Singh SK, Kumar A, Singh BP, Yadav JK, Dubey K. Integrated management of lentil wilt caused by *Fusarium oxysporum* f. sp. *Lentis*. Int J Curr Microbiol App Sci. 2017b;6:1319–22.

Singh R, Singh MK, Singh AK, Singh C. Pulses production in India: issues and elucidations. Pharma Innov. 2018;7:10–3.

Sinha P, Rizvi G, Parashar R. Management of wilt disease of pulses: a review. Int J Pure App Biosci. 2018;6:696–708. https://doi.org/10.18782/2320-7051.6726.

Smitha KP, Rajeswari E, Alice D, Raguchander T. Assesment of vascular wilt and dry root rot of pigeonpea in Tamil Nadu. Int J Tropical Agri. 2015;33:2145–51.

Sousa MVD, Machado JDC, Simmons HE, Munkvold GP. Real-time quantitative PCR assays for the rapid detection and quantification of *Fusarium oxysporum* f. sp. *phaseoli* in *Phaseolus vulgaris* (common bean) seeds. Plant Pathol. 2014;64:478. https://doi.org/10.1111/ppa.12257.

Srivastavaa SK, Sivaramaneb N, Mathura VC. Diagnosis of pulses performance of India. AERR. 2010;23:137–48.

Su G, Suh SO, Schneider RW, Russin JS. Host specialization in the charcoal rot fungus, *Macrophomina phaseolina*. Phytopathology. 2001;91:120–6.

Sunkad G, Sharma M, Mallesh SB, Mannur DM, Sreenivas AG. Distribution and severity of dry root rot of chickpea caused by *Rhizoctonia bataticola* in parts of North Karnataka, India. Int J Curr Microbiol App Sci. 2018;7(4):194–200.

Surulirajan, Tripathi UK, Patel D, Jha DK. Root rot disease of lentil caused by *Sclerotium rolfsii* Sacc. Progressive Res. 2007;2:102–4.

Tetali S, Karpagavalli S, Pavani SL. Management of dry root rot of black gram caused by *Macrophomina phaseolina* (Tassi) using Bio agent. Plant Arch. 2015;15:647–50.

Thakur BR, Kumari N, Singh A. Occurrence of pea root rot/wilt complex disease in Himachal Pradesh. HJAR. 2016;42(2):187–91.

Thokala P, Kamil D, Pandey P, Narayanasamy P, Mathur N. Combined approach of morphological and molecular diagnosis of *Fusaria* spp. causing diseases in crop plants. In: Awasthi LP, editor. Recent advances in the diagnosis and management of plant diseases. New Delhi: Springer; 2015.

Tiwari N, Ahmed S, Kumar S, Sarker A. Fusarium wilt: a killer disease of lentil. Intechopen. 2018:119–38. https://doi.org/10.5772/intechopen.

Tomlinson J, Boonham N. Potential of LAMP for detection of plant pathogens. Perspect Agric Vet Sci Nutr Nat Resour. 2008;3:1–7.

Trivedi S, Srivastava M, Srivastava AK, Ratan V, Shahid M, Singh A, Pandey S, et al. Status of root and foliar fungal diseases of pulses at different agro-climatic zones of Uttar Pradesh, India. Int J Curr Microbiol App Sci. 2017;6:152–65. https://doi.org/10.20546/ijcmas.2017.611.020.

Tsedaley B. A review on disease detection, pathogen identification and population genetics in fungi. J Biol Agric Health. 2015;5:6–20.

Vandemark GJ, Grunwald NJ. Use of real-time PCR to examine the relationship between disease severity in pea and *Aphanomyces euteiches* DNA content in roots. Eur J Plant Pathol. 2005;111:309–16.

Varshney RK, Mohan SM, Gaur PM, Chamarthi SK, Singh VK, Srinivasan S, et al. Marker-assisted backcrossing to introgress resistance to Fusarium wilt race 1 and Ascochyta blight in C 214, an elite cultivar of chickpea. Plant Genome. 2014;7 https://doi.org/10.3835/plantgenome2013.10.0035.

Veena DR, Priya HR, Khatib RM, Joythi D. Soilborne diseases in crop plants and their management. Res Rev J Agri Allied Sci. 2014;3:12–8.
Weller DM. Biological control of soilborneplant pathogens in the rhizosphere with bacteria. Annu Rev Phytopathol. 1988;26:379–407. https://doi.org/10.1146/annurev.py.26.090188.002115.
Winter P, Kahl G. Molecular marker technologies for plant improvement. World J Microbiol Biotechnol. 1995;11:438–48.
Yang A, Juzwik J. Use of nested and real-time PCR for the detection of *Ceratocystis fagacearum* in the sapwood of diseased oak species in Minnesota. Plant Dis. 2017;101:480–6. https://doi.org/10.1094/PDIS-07-16-0990-RE.
Yaqu F, Shahzad S. Effect of solar heating by polyethylene mulching on sclerotial viability and pathogenicity of *Sclerotium rolfsii* on mungbean and sunflower. Pak J Bot. 2009;41:3199–205.

Chapter 7
Diversity of *Phytophthora* Stem Blight of Pigeonpea and Its Sustainable Management

S. J. Satheesh Naik, Abhishek Bohra, T. Basavaraja, R. K. Mishra, G. Padmaja, and K. N. Poornima

7.1 Introduction

Pigeonpea (*Cajanus cajan* (L.) Millisp.) is called by different vernacular names (arhar, tur, redgram, togari, kandalu, etc.), and it is an economically important grain legume of the small and marginal farmers in India. Pigeonpea is one of the major and inseparable dietary protein sources to the large mass of the Indian population (Varshney et al. 2010). Pigeonpea is cultivated as a sole crop and intercrop with rainfed cereals, millets, oils seeds, and other pulses; thereby, it enhances the system productivity and net income to the small and marginal farmers. The differences in the maturity duration of pigeonpea allow it to grow in diversified cropping systems and patterns in varied agro-eco regions of the country.

This has been a matter of concern since the per capita protein availability in India is declining steadily from 27.30 kg/year in 1950 to 10 kg/year in 2009 (Saxena et al. 2014). At present, the national harvest accounts for about 4.25 million tonnes of pigeonpea grains (http://agricoop.gov.in). However, this quantity is not sufficient to meet the domestic needs; about 0.41 million tonnes of pigeonpea is imported annually. The prevailing situation is not likely to improve in the near future by considering the 1.1% annual growth in population (World Bank 2017), plateau of pulse production, inherent low genetic variability for high yield and its attributing traits among the cultivars used in breeding programme and susceptibility of pigeonpea to major

S. J. Satheesh Naik (✉) · A. Bohra · T. Basavaraja
Crop Improvement Division, ICAR-Indian Institute of Pulses Research, Kanpur, India

R. K. Mishra
Crop Protection Division, ICAR-Indian Institute of Pulses Research, Kanpur, India

G. Padmaja
Plant Pathology, PJTSAU, RARS, Warangal, India

K. N. Poornima
Plant Biotechnology Division, ICAR-Indian Institute of Pulses Research, Kanpur, India

B. P. Singh et al. (eds.), *Management of Fungal Pathogens in Pulses*, Fungal Biology, https://doi.org/10.1007/978-3-030-35947-8_7

diseases and insect pests (Ariyanayagam et al. 1995; Yang et al. 2006; Mallikarjuna et al. 2007; Naik Satheesh et al. 2012; Bohra et al. 2014a; Mishra et al. 2016). This opens the new avenue to use the elite genotypes and wild species into the breeding program to create unexplored genetic variability in pigeonpea through pre-breeding (Sharma and Upadhyaya 2016; Saxena and Kumar 2003; Saxena et al. 2010).

In India, the majority of the pigeonpea production comes from states like Madhya Pradesh, Maharashtra, Gujarat, Karnataka, Andhra Pradesh, Telangana, and Uttar Pradesh. In these states, medium- and long-duration pigeonpea cultivars are grown as intercrop, and it is unlikely that the cultivated pigeonpea area will increase by any significant extent to meet the entire need of the country. Hence, new production niches with early-maturing cultivars were explored. As a follow-up pigeonpea, wheat rotation was successfully introduced in the states of Punjab, Haryana, and Western Uttar Pradesh. However, the new varieties which are resistant to *Phytophthora* stem blight disease and photothermal insensitive, a major production constraint, are being marketed through local agro-dealers (Varshney et al. 2014).

The diverse growing conditions expose the pigeonpea to different biotic and abiotic stresses during its life cycle. Pigeonpea get infected by different diseases and insect pests; however, few of them only cause considerable economic losses (Nene et al. 1996; Dhar et al. 2004). After wilt (C.O: *Fusarium udum*) and sterility mosaic disease (SMD) (C.O: Pigeonpea Sterility Mosaic Virus), *Phytophthora* stem blight (PSB) caused by *Phytophthora drechsleri* Tucker f. sp. *cajani* is the third most important disease of pigeonpea in India (Kannaiyan et al. 1984; Mishra et al. 2016) causing complete crop loss upon its infection. PSB has also been reported as the most important production constraint in northeastern states of India (Mishra and Shukla 1987; Chauhan et al. 2002).

7.2 Economic Importance of *Phytophthora drechsleri* Tucker f. sp. *cajani*

The fungus, *Phytophthora drechsleri*, attacks to young (1–7-week-old) plants of pigeonpea, which in turn kills the young plants at the early stage of crop stand to leave large gaps in plant stands (Fig. 7.1). Yield losses are generally higher in early maturing pigeonpea in comparison to medium- and long-duration varieties, because of favorable disease triangle components in early pigeonpea.

7.3 Disease Epidemiology

The *Phytophthora drechsleri* Tucker f. sp. *cajani* survives in soil and infected plant parts as chlamydospores, oospores, and dormant mycelium. Chlamydospore is thick-walled long-term survival spores, as they are produced through asexual means of reproduction. Whereas oospores are sexual spores, these are produced from

Fig. 7.1 *Phytophthora* stem blight infected field of pigeonpea at the early stage (**a**) and later stage (**b**) leaving the large gap in the plant stand

fertilization of the oogonium by an antheridium. Mycelium of *Phytophthora* is coenocytic, aseptate, hyaline, and profusely branching mainly of monopodial branches. The septa are formed at the time of reproduction.

For a successful disease triangle, moist cloudy conditions with drizzling rain are prerequisite, and temperatures between 25 and 28 °C favor rapid infections in young seedlings. The infection requires continuous wetness of plants for about 8 hours to start. As plants grow older, they gradually develop tolerance/resistance to the disease incidence, and they are generally not infected after they are 60 days old. The PSB infection occurs more in organic matter-enriched clay soil in comparison to clayey soil with little organic matter. The disease symptom appears first in low-lying areas of the field where water stagnates. High-density planting, coupled with low availability of resistant varieties, leads to enhanced PSB buildup in early maturating pigeonpea. Warm and humid conditions followed by start-up of an infection of PSB would result in rapid disease development and eventually lead to plant death. Further, speedy wind and rain splashes help to disseminate zoospores. *Phytophthora drechsleri* Tucker f. sp. *cajani* lives on different wild hosts of pigeonpea, for instance, *Cajanus scarabaeoides* var. *scarabaeoides*, a wild relative of pigeonpea, act as a collateral host for *drechsleri* Tucker f. sp. *cajani*.

7.4 Disease Symptoms and Progress of Disease on Pigeonpea

Phytophthora drechsleri present symptomless in the rhizosphere of pigeonpea, and the infection was only evident when the favorable disease triangle exists (Stanier et al. 1971; Lewis 1973). The symptoms of the *Phytophthora* blight disease on pigeonpea have been described in detail by Pal et al. (1970) as stem rot, by Williams et al. (1975) as stem blight, and by Kaiser and Melendez (1978) as a stem canker. The most commonly preferred name for *Phytophthora* infection is the term blight to describe the disease; because all aboveground parts of the pigeonpea plant are affected, further the roots of diseased plants show no symptoms until the plant dies.

Sarkar (1988) reported that the development of PSB is positively correlated with its soil inoculum potential. Bisht (1985) and Sharma et al. (2015) found that zoospores are the primary source of inoculums. Speedy wind helps in spore dispersal over short distances during rain splash. Williams et al. (1975) found high disease incidence due to poor soil surface drainage; in contrary Singh and Chauhan (1985) reported PSB developing to an epidemic level in well-drained fields. Therefore, drainage alone is not the deciding factor for PSB epidemics. Further, Sharma et al. (2006) reported an outbreak of PSB in well-drained, partially drained, and temporarily waterlogged fields irrespective of cropping systems, soil types, and crop cultivars in the Deccan Plateau of India.

Phytophthora stem blight resembles damping off disease at the early stage of infection that causes young seedlings to die after infection. Further infected plants have water-soaked lesions on their leaves and brown to black spots, slightly sunken lesions on their stems and petioles. Infected plant parts lose turgidity and become desiccated. Lesions strap the affected main stem or a branch which leads to break at that infected point, causing the foliage above the lesion to dry up and lodging. Pigeonpea plants that are infected by blight, but not killed, often produce large galls on their stems especially at the edges of the lesions (Fig. 7.2).

Singh and Chauhan (1985) reported more rapid development of PSB at night in the field due to favorable disease development conditions; this hypothesis was confirmed under artificial darkness conditions in the greenhouse. Reddy et al. (1991a, b) confirmed the PSB infection usually occurs when there is a decrease in day temperatures of the previous week, and the difference between the maximum and minimum temperatures are the least. Studies on relationships between PSB incidence and soil nutrition indicated that in the absence of potassium (K) and high doses of nitrogen (N), PSB incidence increased (Pal and Grewal 1975). Nevertheless, the addition of K decreased disease incidence regardless of the presence of N or phosphorus (P) in the soil (Fig. 7.3).

Fig. 7.2 Phytophthora infected pigeonpea plants at the early stage (**a**) and later stage with large galls on the stems (**b**)

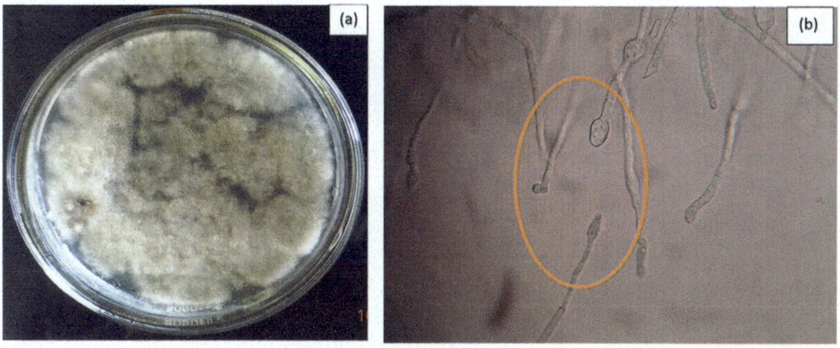

Fig. 7.3 (**a**) Cottony mycelial growth of PSB on V8 juice agar. (**b**) The hypal structure and 40× magnified papiliate hybphae of PSB

7.5 Morphological Features of *Phytophthora*

The cell wall of *Phytophthora* is made up of cellulose. *Phytophthora drechsleri* Tucker f. sp. *cajani* resembles true fungi because they grow using fine filaments called hyphae and produce spores. *Phytophthora* hyphae lack cross wall septa and diploid phase. The *Phytophthora drechsleri* Tucker f. sp. *cajani* has terminal papillate hyphae which in turn produces the spores. The sizes of sporangia of *Phytophthora drechsleri* var. *cajani* ranging from 42 to 83 × 29 to 48 μm (average 61.8 × 37.3 μm) and the sporangial stalks is either narrowly tapered or widened somewhat at the base of the sporangium (Fig. 7.4b).

Phytophthora produces several types of substructure that are specialized for survival during the adverse condition of their life cycle. Chlamydospores and oospores are prominent spores of *Phytophthora* produced during the adverse conditions of their growth and development. Chlamydospores are thick-walled long-term survival spores produced by asexual means of reproduction, while oospores are sexual spores, which are produced from fertilization of the oogonium and antheridium.

7.6 Disease Management Techniques

In any disease management, host plant resistance is the primary step for exploring available germplasm stocks and breeding lines to identify donors. Different techniques for PSB resistance screening under field and greenhouse conditions have been reported by various researchers. Pal et al. (1970) used a "leaf scar" method to inoculate 30- to 60-day-old seedlings which are grown in pots under greenhouse conditions. This method consisted of inoculating plants at the point of attachment of leaf after its removal with mycelial mats of the fungus multiplied on potato dextrose agar. Kannaiyan et al. (1981) standardized the pot-culture drench inoculation and foliage inoculation techniques. In drench inoculation, 5- to 10-day-old seedlings raised in pots filled with sterilized field soil are drench-inoculated with the

Fig. 7.4 (**a**) Ridge planting of pigeonpea at early seedling stage. (**b**) Established pigeonpea crop on ridge planting method

macerated mycelial suspension of the fungus multiplied on V-8 juice medium (one mycelial mat in 200 ml of water). Inoculum (100 ml) was poured around seedlings. Pots were liberally watered three times a day to assure adequate development of the disease. In this technique, the disease developed after 7–10 days of inoculation. In the foliage inoculation technique, the inoculum is sprayed on 15- to 30-day-old plants grown in a pot, the plants covered with polythene bags for 48 h, kept on glasshouse benches, and later sprayed with water for 10 days. Typical blight symptoms appeared within 10 days after the inoculations.

The sick field screening of pigeonpea genotypes for *Phytophthora* blight resistance was standardized at ICRISAT and ICAR-IIPR, Kanpur, including planting of test entries with 30 cm row spacing and interplanting a susceptible cultivar (e.g., ICP 2376, UPAS 120, ICP 1134, and ICP 7119) to serve as an indicator line after every 2–4 rows. The sick field was prepared by incorporating diseased debris of susceptible cultivars; further, the inoculum load in the sick field is maintained through periodical soil sample analysis of PSB sick field. Additional sickness in the field is created by incorporating infected plant debris.

Agronomic intervention plays an important part in the management of PSB disease. The desiccation of pathogenic spore and dormant mycelium through summer solarization or summer ploughing of field is being done to avoid the inoculum load. Practicing the ridge planting method is highly advantageous to drain excess rainwater since pigeonpea requires well-aerated soil for its growth and development. After the onset of monsoon, timely sowing is highly advisable for establishing early growth and in turn keeping away the disease incidence, because older plants are more resistant to *Phytophthora* blight disease due to systemic acquired resistance. Select fields with no previous record of PSB, and avoid sowing pigeonpea in fields with low-lying patches that are prone to temporary waterlogging. Use wide inter-row spacing for good aeration and plant growth.

Although several fungicides have proved effective in the control of PSB, however, systematic studies on the control of soilborne diseases like PSB using fungicides are limited. In a pot experiment, Pal and Grewal (1983) reported Brestan-60 effective in controlling PSB in 1-month-old plants when applied before inoculating with PDC. Significant control of blight (>90%) was achieved with metalaxyl (1.75 g a.i kg^1 seed) in a greenhouse experiment (Agarwal 1987; Bisht and Nene 1988). However, Chaube et al. (1987) reported the poor efficacy of metalaxyl applied as a seed dressing in protecting older pigeonpea plants against PSB. At the later stage of PSB, the infection plant develops galls and makes them susceptible to lodging during intercultural operation and speedy wind. Sheila and Nene (1987) reported reduced PSB incidence with the spray or soil drench with two phytoalexins like Phytoalexin-84 and Induce. Park et al. (2007) claim that the direct application of slow-releasing phosphorous acid formulations (curdlan or pestan) using a carrier coated with polysaccharides resulted in an excellent control of PSB disease of pepper. They further suggested that the application of formulation product once or twice during crop season can control *Phytophthora* diseases on various crops. However, there is no evidence in pigeonpea to say this product can be used for the management of PSB in pigeonpea.

Practicing of the integrated disease management (IDM) technology is essential for economical and sustainable means to control PSB. Moderate levels of host plant resistance-bred varieties can be combined with other cultural practices, and application of minimal dosage of fungicide for control of PSB would save large input cost to farmers. The recommended IDM practices include (a) use of pathogen-free seed, (b) seed treatment with fungicide, (c) crop rotation, (d) raised bed planting, (e) adequate field drainage, and (f) use of disease resistant variety, and strategic application of fungicides will help in the management of disease in a sustainable manner.

7.7 Future Prospective and Conclusion

Phytophthora blight (*Phytophthora drechsleri* f. sp. *cajani*) is one of the major yield limiting factors of short-duration varieties of pigeonpea (*Cajanus cajan*). For eco-friendly and sustainable management of the disease, antagonists

(*Pseudomonas fluorescens*, *Bacillus subtilis*, *Trichoderma viride*, and *T. hamatum*) were evaluated widely and used as bioagents and can be integrated with fungicides for effective management of PSB disease. Commercially available metalaxyl formulation – Ridomil MZ – is also at a par with apron in respect to efficacy against *P. drechsleri* f. sp. *cajani*, and they could be integrated with *P. fluorescens* and *T. viride* for better and eco-friendly management of *Phytophthora* blight of pigeonpea. Ridomil MZ has an additional advantage that it possesses different modes of action and there is a lower chance of cross-resistance with metalaxyl-resistant populations. Mancozeb in combination with metalaxyl was found to be highly effective at reducing disease. However, the chemical method of controlling PSB is not economical and eco-friendly. Therefore more focus is needed for the development of resistant varieties for sustainable management and for higher productivity per unit area.

Acknowledgments We are thankful to the Director, ICAR-IIPR, Kanpur, for extending facilities.

Bibliography

Agarwal SC. Fungicidal control of stem blight of pigeonpea caused by Phytophthora drechsleri Tucker f.sp. cajani. Indian J Plant Prot. 1987;15:35–7.

AICRP on pigeonpea, project coordinators report 2015. IIPR, Kanpur.

Ariyanayagam RP, Nageshwara A, Zaveri PP. Cytoplasmic genic male sterility in interspecific matings of pigeonpea. Crop Sci. 1995;35:981–5.

Bisht VS. Phytophthora blight recent studies. Pigeonpea Pathology Progress Report, ICRISAT, Patancheru, AP, India: Legumes Program; 1985. 44pp.

Bisht VS, Nene YL. A selective medium for Phytophthora drechsleri f. sp. cajani causing pigeonpea blight. Int Pigeonpea Newslett. 1988;8:12–3.

Bohra A, Saxena RK, Saxena KB, Sameerkumar CV, Varshney RK. Advances in pigeonpea genomics. In: Gupta S, Nadarajan N, Sen Gupta D, editors. Legumes in the Omic Era: Springer, New York, Heidelberg Dordrecht, London, 2014a. p. 95–110.

Bohra A, Singh IP, Yadav AK, Pathak A, Soren KR, Chaturvedi SK, et al. The utility of informative SSR markers in the molecular characterization of cytoplasmic genetic male sterility-based hybrid and its parents in pigeonpea. Natl Acad Sci Lett. 2014b;38:13–9.

Chaube HS, Razdan VK, Singh US. Effect of metalaxyl on growth, sporulation and sporangial germination of *Phytophthora drechsleri* f. sp. *cajani*. Int Pigeonpea News lett. 1987;6:59–61.

Chauhan VB, Singh VB, Singh AK. Status of Phytophthora blight of pigeonpea in eastern Uttar Pradesh. Ann Pl Protec Sci. 2002;10:402–4.

Dhar V, Singh RA, Gurha SN. Integrated disease management in pulse crops. In: Masood A, Singh BB, Kumar S, Dhar V, editors. Pulses in new perspective. Kanpur, India: Indian Society of Pulses Research and Development, IIPR; 2004. p. 325–44.

Dhar V, Reddy MV, Chaudhary RG. Major diseases of pigeonpea and their management. In: Masood A, Kumar S, editors. Advances in pigeonpea research; 2005. p. 229–61.

Gooding HJ. The agronomic aspects of pigeonpeas. Field Crop Abstracts. 1962;15:1–5.

Gupta AK, Singh IS, Reddy MV, Bajpai GC. Genetics of resistance to P3 isolate of Phytophthora blight in pigeonpea. Euphytica. 1997;95:73–6.

Kaiser WJ, Melendez PLA. Phytophthora stem canker disease of pigeonpea in Puerto. Rico PI Dis Rep. 1978;62:240–2.

Kannaiyan J, Nene YL, Raju TN, Shiela VK. Screening for resistance to Phytophthora blight of pigeon pea. Plant Dis. 1981;65:61–2.

Kannaiyan J, Nene YL, Reddy MV, Ryan JG, Raju TN. Prevalence of pigeonpea diseases and associated crop losses in Asia, Africa and Americas. Trop Pest Management. 1984;30:62–71.

Lewis DH. Concepts in fungal nutrition and the origin of biotrophy. Biol Rev. 1973;48:261–78.

Mallikarjuna N, Jadhav D, Reddy MV, Dutta TU. Introgression of Phytophthora blight disease resistance from Cajanus platycarpus into short duration pigeonpea [Cajanus cajan (L.) Millsp.]. Indian J Genet. 2005;65:261–3.

Mallikarjuna N, Sharma HC, Upadhyaya HD. Exploitation of wild relatives of pigeonpea and chickpea for resistance to Helicoverpa armigera. SAT eJ. 2007;3:1–4.

Mallikarjuna N, Jadhav DR, Srikant S, Saxena KB. *Cajanus platycarpus* (Benth.) Maesen as the donor of new pigeonpea cytoplasmic male sterile (CMS) system. Euphytica. 2011;182:65–71.

Mishra AN, Shukla P. Prevalence of Phytophthora blight of pigeonpea in Uttar Pradesh. Indian Phytopath. 1987;40:56–8.

Mishra RK, Naimuddin, Saabale PR, Naik Satheesh SJ, Krishna K, Singh F, Singh IP. Evaluation of promising lines of pigeonpea for resistance to wilt caused by *Fusarium udum* Butler. J Food Legume. 2016;29(1):64–6.

Naik Satheesh SJ, Byre Gowda M, Venkatesha SC, Ramappa HK, Pramila CK, Marry Reena GA, Ramesh S. Molecular diversity among Pigeonpea genotypes differing in response to Pigeonpea sterility mosaic disease. J Food Legumes. 2012;25(3):194–9.

Nene YL, Sheila VK, Sharma SB. A world list of chickpea and Pigeonpea pathogens. 5th ed. Patancheru, India: ICRISAT; 1996.. 27pp

Pal M, Grewal JS. Utilization of different nitrogen sources by Phytophthora drechsleri var. cajani. Indian Phytopathol. 1975;28(4):499–501.

Pal M, Grewal JS. Chemical control of Phytophthora blight of pigeonpea. Indian Phytopathol. 1983;36:380–1.

Pal M, Grewal JS, Sarbhoy AK. A new stem rot of arhar caused by Phytophthora. Indian Phytopathol. 1970;23:583–7.

Park HJ, Kim SH, Jee HJ. A new formulation system for releasing of phosphorous acid in soil for controlling Phytophthora diseases. Plant Pathol J. 2007;23:26–30.

Reddy MV, Nene YL, Raju TN, Sheila VK, Sarkar N, Remanandan P, Amin KS. Pigeonpea lines field-resistant to Phytophthora blight. Int Pigeonpea Newslett. 1991a;13:20–2.

Reddy MV, Sarkar N, Nene YL, Raju TN. Predisposing factors for Phytophthora blight of pigeon-pea. Indian Phytopathol. 1991b:268–70.

Reddy MV, Raju TN, Sheila VK. Phytophthora blight disease in wild pigeonpea. Int Chickpea Pigeonpea Newslett. 1996;3:52–3.

Sameer Kumar CV, Singh IP, Suyash BP, Myer GM, Kumar VR, Saxena RK, Varshney RK. Recent advances in Pigeonpea [*Cajanus cajan* (L.) Millspaugh) Research, In: II International Conference on Bio-Resource and Stress Management, January 07–10, 2015, Hyderabad; 2015.

Sarkar N. Epidemiological studies on Phytophthora blight of pigeonpea. Pulse pathology Progress report 53. Patancheru, AP, India: Legumes Program ICRISAT; 1988. 17pp.

Saxena KB, Kumar RV. Development of cytoplasmic nuclear male-sterility system in pigeonpea using C. scarabaeoides (L.) Thours. Indian J Genet. 2003;63:225–9.

Saxena KB, Kumar RV, Dalvi VA, Mallikarjuna N, Gowda CLL, Singh BB, et al. Hybrid breeding in grain legumes: a success story of pigeonpea. In: Khairwal MC, Jain HK, editors. Proceedings of the International Food Legumes Research Conference. New Delhi; 2005.

Saxena KB, Sultana R, Mallikarjuna N, Saxena RK, Kumar RV, Sawargaonkar SL, et al. Male-sterility systems in pigeonpea and their role in enhancing yield. Plant Breed. 2010;129(2):125–34.

Saxena KB, Singh IP, Kumar RV, Hingane AJ, Mula MG, Patil SB, Kumar CVS. Challenges and opportunities of breeding early maturing pigeonpea hybrids. J Food Legumes. 2014;27(1):1–8.

Sharma S, Upadhyaya HD. Pre-breeding to expand primary genepool through introgression of genes from wild Cajanus species for pigeonpea improvement. Legume Perspectives. 2016;11:17–20.

Sharma M, Pande S, Pathak M, Narayana RJ, Anilkumar P, Reddy M, Benagi D, Mahalinga VI, Zhote DM, Karanjkar KK, Eksinghe PN. Prevalence of Phytophthora blight of pigeonpea in the Deccan Plateau in India. Plant Pathol J. 2006;22:309–13.

Sharma M, Ghosh R, Tarafdar A, Telangre R. An efficient method for zoospore production, infection and real-time quantification of Phytophthora cajani causing Phytophthora blight disease in pigeonpea under elevated atmospheric CO2. BMC Plant Biol. 2015;15:1–12. https://doi.org/10.1186/s12870-015-0470-0.

Sheila VK, Nene YL. Efficacy of Phytoalexin Formulations against Phytophthora drechsleri f.sp. cajani. Int Pigeonpea Newslett. 1987;6:61–2.

Singh UP, Chauhan VB. Relationship between filed levels and light and darkness on the development of Phytophthora blight of pigeonpea (*Cajanus cajan* (L.) Millsp.). Phytopathol Z. 1985;114:160–7.

Spence JA, Williams SJA. Use of photoperiod response to change plant design. Crop Sci. 1972;12:121–2.

Stanier RY, Doudoroff M, Addberg EA. General microbiology. 3rd ed. London: Macmillian; 1971.

Subbarao GV, Johansen C, Kumar RJVDK, Jana MK. Salinity tolerance in F1 hybrids of pigeonpea and a tolerant wild relative. Crop Sci. 1990;30:785–8.

Tikka SBS, Parmar LD, Chauhan RM. First record of cytoplasmic-genic male sterility system in pigeonpea (*Cajanus cajan* (L.) Millsp.) through wide hybridization. Gujarat Agric Univ Res J. 1997;22:160–2.

Upadhyaya HD, Reddy KN, Sastry DVSSR, Gowda CLL. Identification of photoperiod insensitive sources in the world collection of pigeonpea at ICRISAT. SAT eJ. 2007;3(1):1–4.

Vales MI, Srivastava RK, Sultana R, Singh S, Singh I, Singh G, Patil SB, Saxena KB. Breeding for earliness in pigeonpea: development of new determinate and nondeterminate lines. Crop Sci. 2012;52:2507–16.

Van der Maesen LJG. Pigeonpea: origin, history, evolution and taxonomy. In: Nene YL, Hall SD, Sheila VK, editors. The pigeonpea. Wallingford: CAB International; 1990. p. 15–46.

Varshney RK, Penmetsa RV, Dutta S, Kulwal PL, Saxena RK, Datta S, et al. Pigeonpea genomics initiative (PGI): an international effort to improve crop productivity of pigeonpea (*Cajanus cajan* L.). Mol Breed. 2010;26:393–408.

Varshney RK, Terauchi R, Mc Couch SR. Harvesting the promising fruits of genomics: applying genome sequencing technologies to crop breeding. PLoS Biol. 2014;2:100–883.

Wanjari KB, Patil AN, Manapure P, Manjaya JG, Manish P. Cytoplasmic male-sterility with cytoplasm from Cajanus volubilis. Ann Plant Physiol. 2001;13:170–4.

Williams FJ, Amin KS, Baldev B. Phytophthora stem blight of Cajanus cajan. Phytopathology. 1975;65:1029–30.

World Bank Annual Report 2017. Washington, DC: World Bank. https://doi.org/10.1596/978-1-4648-1119-7

WWW.agricoop.nic.in/imagedefault1/Pulses.pdf

WWW.agricoop.nic.in/site/default/files/3rdAdv150216Eng.pdf

Yang S, Pang W, Harper J, Carling J, Wenzl P, Huttner E, et al. Low level of genetic diversity in cultivated pigeonpea compared to its wild relatives is revealed by diversity arrays technology (DArT). Theor Appl Genet. 2006;113:585–95.

Chapter 8
Foliar Fungal Diseases in Pulses: Review and Management

Priyanka Kakoti, Parishmita Gogoi, Archana Yadav, Bhim Pratap Singh, and Ratul Saikia

8.1 Introduction

Pulses play a major role in the nutritional security of people having a cereal-based diet. The United Nations Food and Agriculture Organization (FAO) recognizes 11 types of pulses in India as chickpea (brown and green), lentil (masoor), faba bean (broad beans), field pea (matar), cowpeas, black gram, black-eyed bean, pigeonpea (arhar), and red kidney beans (rajma) (Busby et al. 2016). Besides their value as diet and having nitrogen fixation ability, pulses also play an important role in flourishing intensive agriculture by improving the physicochemical and biological properties of soil. Aerial fungi attack causes diseases like gray and chocolate spots, *Ascochyta* blight, anthracnose, leaf rot, powdery mildew, leaf yellowing, stem canker, and downy mildew. These diseases are caused by a fungus that can be necrotrophic or biotrophic, e.g., *Botrytis cinerea*, *B. fabae*, *Ascochyta rabiei*, *Colletotrichum*, *Puccinia triticina*, *Erysiphe polygon*, and *Perenospora* (Trivedi et al. 2017). However, some of them affect larger areas among several countries where the cultivation of legumes occurs and cause degradation in quantity and quality.

Development of disease by fungi in host plants is a stepwise phenomenon, starting from the contamination phase, following contact between the host plant and spores of the fungus. Depending on adequate receptivity and compatibility, spore germination occurs and forms appressoria that allow the fungus to penetrate directly the host plant or by cuticle, stomata, or tissues wounded. Infection follows penetration, where the fungus settles and invades the host tissue, enhancing its

P. Kakoti · P. Gogoi · A. Yadav · R. Saikia (✉)
Biotechnology Group, Biological Science & Technology Division,
CSIR-North East Institute of Science and Technology, Jorhat, Assam, India
e-mail: ratulsaikia@neist.res.in

B. P. Singh
Department of Biotechnology, Mizoram University, Aizawl, Mizoram, India

B. P. Singh et al. (eds.), *Management of Fungal Pathogens in Pulses*,
Fungal Biology, https://doi.org/10.1007/978-3-030-35947-8_8

development either on a dead (necrotrophic fungus) or on living tissue (biotrophic fungus), resulting in the development of symptom (Sinha et al. 2018). The fungi then develop specialized structures to carry out the production of secondary inoculum by sporulation that contributes to starting another infection cycle. Measures for controlling these diseases including identification of resistant germplasm, choosing varieties resistant to fungus by screening and experimentation, cultural management, chemicals, genetic resistance, or combination of such approaches are required, and attention has been given in this direction by the researchers (Pautasso et al. 2012).

8.2 Fungi Affecting Foliar Parts

Pulses are consumed as a chief source of plant protein. Consisting of amino acids, they have medicinal properties as well. Pulse crops are cultivated during rabi, zaid, and kharif seasons of the agricultural year. Rabi crops require mild cold climate during sowing to pod development and warm climate during maturity/harvesting, e.g., chickpea, lobia, moong, pigeonpea, urad, masur, etc., affected by fungi such as *Colletotrichum* (Dilani et al. 2017), *Uromyces*, *Cylindrosporium*, etc., whereas kharif pulse crops require warm climate throughout their life that is from sowing to harvesting, e.g., arhar, black gram, cowpea, moong, and urad, usually attacked by *Erysiphe*, *Cercospora*, *Fusarium*, *Ascochyta*, *Alternaria*, *Phoma*, etc. (Tivoli et al. 2006). In most of the fungal groups, the temperature varies according to species. Some diseases and their common causal organisms are cited in Table 8.1.

Table 8.1 Some pulse diseases and their causal agents

Disease/ symptom	Causal agent	References
Blight	*Alternaria alternate*, *Ascochyta fabae*, *Stemphylium botryosum*	Akem (1999), Davidson and Kimber (2007)
Anthracnose	*Colletotrichum truncatum*	Kim et al. (2015), Than et al. (2008)
Leaf spot	*Cercospora lentis*, *C. cruenta*, *C. zonata*, *Cylindrosporium* sp., *Helminthosporium*, *Phoma medicaginis*, *Pestalotia* sp.	Suterman et al. (2011), Ringer and Grybauskas (1995)
Stem canker	*Cylindrosporium* sp.	Nikmaram et al. (2017)
Downy mildew	*Perenospora lentis*, *P. viciae*	Madden et al. (2007), Farouk et al. (2017), Xin et al. (2011)
Wilt	*Fusarium oxysporum*	Oumouloud et al. (2013)
Leaf rot	*Choanephora* sp.	Gossen et al. (2016)
Leaf yellowing	*Cladosporium herbarum*, *Pyrenophora tritici-repentis*	Raimondo and Carlucci (2018)
Powdery mildew	*Erysiphe polygoni*, *Podosphaera xanthii*	Sparks and Kelly (2017), Caffarra et al. (2012)

8.3 Overview of Common Foliar Diseases of Pulses

8.3.1 Blight Disease

Blight can be considered as complete chlorosis, which includes browning and death of plant tissues such as leaves, twigs, branches, and floral organs and fruits. Blight mostly appears as water-soaking spots, toward the edge of lower leaves where dew or water gets collected (Davidson and Kimber 2007). Under cool and moist conditions, water-soaking spots enlarge faster, and a yellow broad portion might be seen around the lesion. While on the underneath, white mold growing zone producing spore (approx. 0.1–0.2 inches wide) may appear at the lesion border. Under a wet environment, disease flourishes faster. Dry and warm weather slow down or stop disease development; however, it resumes with weather conditions being moist again (Akem 1999). Spores are readily disseminated by rain splashing, over-irrigation, or wind. Repeating cycles of production of spores, dissemination, and production of extra spores give blight disease its explosive potential. Late blight is most aggressive among all because of its polycyclic nature as it goes through several disease cycles within a year. Seven fungicides were evaluated in vitro against *Exserohilum turcicum* that causes leaf blight (Reddy et al. 2013). The mancozeb (0.25%) alone or combination with carbendazim reduced the disease up to 72–73%.

8.3.2 Anthracnose

This fungal disease mostly attacks plants during the spring with cool and wet weather, on leaves and twigs. Cool, rainy weather creates favorable conditions for the spores to spread. Fruiting bodies appear as tiny dispersed black-colored flecks, and pink masses of spores are seen at the center of the old black spot.

Colletotrichum uses different strategies to cause infection of the host plant which starts from the hemibiotrophic intracellular mode up to the necrotrophic nutrition mode (Bailey and Jeger 1992). But different species undergo diverse infection mechanisms depending on the host plant infected. The initial infection starts as the conidia attach to the host surface, germinate, and produce appressoria following penetration of host epidermis. Fungus colonizing plant tissue results in the formation of certain structures called acervuli that contain spores. The pathogen stays in the inert state sometimes in the form of appressoria in tissues of unripe fruits, and infection is caused after it ripens. The management and control of anthracnose diseases are still being studied. Many studies have concluded that disease management practices are often insufficient to eradicate these diseases. Breeding techniques to develop long-lasting resistant varieties are also not successful due to the involvement of multiple *Colletotrichum* species in anthracnose infection (Than et al. 2008). Different species are reported to attack different organs of the host plant, e.g.,

C. acutatum and *C. gloeosporioides* infect fruits of host plant at all developmental stages, and the leaves or stems are mostly damaged by the species such as *Colletotrichum coccodes* and *Colletotrichum dematium* (Kim et al. 2015).

8.3.3 Leaf Spot Disease

Leaf spots, round in shape are found on the leaves of many plant species, mostly caused by fungi that are parasitic in nature. A typical spot has a defined edge and is often dark at the border. When many spots are present, together they can grow and form a blotch or blight. Spots of fungi are usually of free form or round in shape. In spring, when conditions are in favor of the fungus, ascospores discharge from perithecia and infect young leaves of plants. Once infected, a leaf serves as a good nutrient source for the fungus to produce secondary inoculum (conidia) inside pycnidia (the surviving structure that protects the spores). Conidia are capable of undergoing several repeated secondary cycles and re-infect other plants nearby. When the leaves of the plant start falling, asci and ascospores are produced within perithecia and are protected until the following spring. The ascospores are characterized by a cylindrical, curved shape, pointed at both ends with four septa (Ringer and Grybauskas 1995). A temperature of 30 °C is favorable for maximum colony growth and acervuli production. The optimum temperature for growth and sporulation of *Pestalotia* sp. was 25 °C. Germination of spores requires 30 °C, and they don't germinate below 15 °C or above 40 °C (Ramaswamy and Sohi 1984; Naqvi 2004). It was reported that rainfall, relative humidity, and temperature are the weather components significantly affecting the increase of disease severity of *Pestalotia* (Suterman et al. 2011).

8.3.4 Stem Canker

Stem canker is often confused with *Phytophthora* stem rot. Green stem tissue appears below the canker, while it is not present in root rot, there is none. Necrosis and interveinal foliar chlorosis may occur as a result of fungus producing toxins. These symptoms may be similar to those of sudden death syndrome and brown rot. Stem canker is noticed at the latter half of the growing season. During the early reproductive stages of plants, reddish-brown stem lesions appear which are in the portion of the stem node (Backman et al. 1985). The pathogen can survive in the residue of host or the soil for many years in the form of spores which act as the primary source. During rainy weather, spores are produced in the early vegetative stages which splash onto plant tissue causing infection. Until the plant enters the reproductive stage, cankers are not visible on plant tissue where secondary spore production may take place. Infection can occur over temperatures of a wide range, but the fungus needs the moist condition to infect (Nikmaram et al. 2017).

8.3.5 Downy Mildew

Downy mildew is caused by oomycete organism. It is spread by airborne spores. The infection is enhanced by prolonging wetness of leaf. Spore formation can occur within 4 days after the initial infection. However, the typical period for germination of spores is 7–10 days (Madden et al. 2007). There are many downy mildew species capable of spore germination by the creation of a germ tube that enters the host. Some species also germinate by zoospores. However, some downy mildew species cannot handle the cool weather and so are reintroduced to another area for another infection to occur (Vittorio et al. 2007). Chitosan application was significantly superior to other elicitors to increase shoot length, nitrogen and phosphorus percentage, photosynthetic pigment, and ascorbic acid, proline, and phenolic compounds of the leaves (Xin et al. 2011). The silicon reduces downy mildew disease severity (Farouk et al. 2017)

8.3.6 Wilt

Wilt disease affects the vascular system of plants. It starts with vein clearing on younger leaves and dropping the old ones toward the lower side, followed by stunting, defoliation, marginal necrosis, yellowing of leaves toward the lower side, and death of the plant. The most abundant is microconidia. Chlamydospores can survive in the soil for a longer time. The mycelium grows intracellularly into the xylem through the root cortex, exclusively within the vessels, and produces microconidia (Saikia et al. 2004a, b). It enters the stream sap and is upwardly transported and germinates where the flow stops. Eventually, the mycelia and the spores clog the vessels of xylem, which prevents the plant from translocating nutrients and their uptake. In the end, the plant transports less and transpires more resulting in stomatal closure, wilting of leaves, and death of the plant. After the plant's death, the fungus sporulates by invading all tissues and continues infecting other nearby plants. The development and deployment of resistant cultivars are generally considered to be the best approach to control *Fusarium* wilt. Two dominant resistance genes fom-1 and fom-2 play an important role in controlling resistance in various races of the host (Oumouloud et al. 2013).

8.3.7 Leaf Rot

In leaf rot, lesions are water soaking with various colors and shape formed on the appearing spindle and young leaves; thus the leaflet does not open fully. Central shoots are affected, and further, all the crown leaves get rotted (Gossen et al. 2016). The lesions enlarge and fuse leading to extensive rot of spindle leaves. Rotting results in the decay of buds as it extends toward the interior of the spindle which further causes the yellowing of leaves. Further infections of the emerging spindles result in the appearance of symptoms in most of the crown leaves. The pathogen survives as long as debris of the infected plant remains. The remaining debris lying on the soil is often the source for primary inoculation that infects other plants of upcoming seasons. High humidity and moisture (dew) on the leaves are needed for the pathogen to infect the host. *Cercospora zeae-maydis* is an atypical pathogen (Aref and Anderson 1973), whose conidia before penetration can grow and survive for many days. But most spores need to be penetrated within hours of germination for ensuring survival. Considering the weather favoring conditions, the conidia for upper leaf regions may also serve as secondary inoculum. Additionally, heavy rains and wind tend the conidia to disperse during many secondary cycles to other parts of the field causing more secondary infection cycle. In adverse conditions, the pathogen undergoes an interstate and reactivates when conditions are favorable.

8.3.8 Leaf Yellowing

According to the recent report, *P. pauciseptata* and *P. ramiseptata* are the most aggressive species causing leaf yellowing in plants (Raimondo and Carlucci 2018). Yellowing of leaves may be caused by manganese, zinc, or nitrogen deficiencies. It is widely known as chlorosis. The yellow spot of disease, caused by *Pyrenophora tritici-repentis*, is a stubble-borne disease. The fungus survives on stubble in small fruiting black bodies, asci, from season to season. They contain ascospores in large numbers which are in humid conditions ejected forcibly. However, at wet conditions and temperatures between 10 and 25 °C, the second type of spore, conidia, is produced. Disease development in higher plants, pulses, and other crops can occur by the secondary spore. It is one kind of secondary infection that leads to loss of high yield.

8.3.9 Powdery Mildew

Mildew is marked by a white floury covering comprising of conidia. The lower leaves are mostly affected, but it is also seen aboveground part of the plant as well. As the disease progresses, the spots get larger and denser as large numbers of asexual spores are formed, and the mildew spreads on the entire host including pods. All species of powdery mildew fungus require living tissues of plant for growth. They survive on stem and bud tissue in perennials. The optimum temperature between 68 and 77 °F and relative humidity between 40% and 100% are favorable for spore germination. Powdery mildew development is also favored by low, diffuse light (Caffarra et al. 2012). Powdery mildew in pulses (mungbean) is caused by the

Podosphaera xanthii, responsible for yield losses of up to 40% (Sparks and Kelly 2017). The mildew spreads faster as the disease cycle can be finished in about 72 hours. However, it takes 7–10 days from the time of infection to the development of symptoms and the production of secondary spores.

8.4 Management

The major prominence in research and development to mitigate pulse diseases is chemical control and resistance of host against pathogens. Recently, a shift in management practices of pulse diseases is seen, and emphasis was given on identifying, evaluating, and integrating components specific to location for integrated disease management (IDM). In general, IDM follows certain principles (Bailey and Jeger 2000). Single component or in combination of other components (fungicide and seed treatment), are used adequately to mitigate pulse diseases. The major components of IDM are the resistance of host plant, disease modeling for the avoidance of high risk or pressure of disease, use of chemicals, biological control, and cultural agronomic practices (Pandel et al. 2009).

8.4.1 Resistance of Host Plant

The interaction between the pathogen and the host defines race specificity or non-race specificity of resistance and is based on the presence or absence of statistically significant interaction between host and pathogen genotypes. It is hard to identify the clear host-fungus interaction or relationship in nature that entirely fits these definitions. Most plant pathogenic fungi show different interactions with their host plants, changing their relationship at different stages of their life cycle depending on the resistance of the host, physiology, the environment, and associated virulence genes of the pathogens. Each intracellular structure also prevents non-specific defense of plants triggered by activities of fungi, possibly by intrusion with the system signaling rather than the expression of defense. In resistant cultivars of the host plant, rapid cell death is triggered by the cellular invasion that shares some features with apoptosis of plant tissues and is controlled by resistant genes that are parasite-specific which resemble genes that defend plants against other types of pathogens (Oumouloud et al. 2013). Evidence suggests that a fungal peptide elicit this response which does not involve the oxidative burst typical of expression of resistance in other pathogen and plant interactions (Heath 1997). However, in general, few of the molecules involved in any fungi and plant interactions have been characterized completely, and much is left to be discovered (Farouk et al. 2017).

8.4.2 Disease Modeling

Disease models help to understand how the disease develops and approaches to test potential mitigation steps. Plant diseases account for about 16% or more of the total yield losses every year. To forecast the spread of these diseases both locally and over long distances, numerical models and monitoring networks have been developed (Knogge 1996). The epidemics of these airborne diseases depend on the production of infectious propagules, their aerial transport, specific infectiousness, and finally their reproduction (Pan et al. 2010). For modeling disease development, various approaches such as statistical modeling, growth curve modeling, and mechanistic modeling are developed. A common core of disease epidemic models is the relationship between disease intensity (y) and time (t), which is given by dy/dt, e.g., $dy/dt = rL$; rL is the parameter determining how fast the disease develops and is dependent on environmental conditions (Maanen and Xu 2003).

8.4.3 Chemical Control

Chemical fertilizers provide nutrients for healthy plant growth which are a combination of synthetic primary nutrients as nitrogen, potassium, and phosphorus. They provide the benefit of having more nutrients than organic ones. The different types of chemicals used in agriculture are insecticides, herbicides, soil fumigants, desiccants, fungicides, plant growth regulators, and harvest aids because natural pesticides are not enough for conventional agriculture (Meyer et al. 2016). Organic farmers use a wide range of natural pesticides for controlling weeds, insects, and diseases. The benefits of using chemicals include increasing yield potential that allows farmers to farm more acres of land and protects the soil through conservation methods. Chemical fertilizers and pesticide use peaked in the 1980s but are declining as farmers and scientists are inclined to eco-friendly control methods.

8.4.4 Biological Control

Many microbes show antagonistic activity against fungal pathogens which could be used to prepare solid or liquid microbial formulations to apply on sensitive and diseased plants (Passari et al. 2017, 2019). The use of these microbes also helps the plants in developing resistance against the fungal infections, e.g., *Bacillus* sp., *Pseudomonas* sp., *Ochrobacterium* sp., etc., which also helps plant growth promotion (Saikia et al. 2003, 2005, 2018). They also induce a defense mechanism against the pathogens in host plants through induced systemic resistance (ISR) (Saikia et al. 2003, 2006). The antagonistic activity of some microbes showed prominent inhibition against the pathogen. This would be helpful for the detection and control of the devastating disease (Chowdhury et al. 2018). In general biocontrol agents suppress pathogens and other organisms. However, the interrelationships of many environmental factors can result in multiple interactions among organisms and their environment, several of which might contribute to effective biological control. Furthermore, some natural products also lead to the development of biorational pesticides (Passari et al. 2017; Gardener and Fravel 2002). Prospects for using biological control are to limit the damage of plant pathogens in both conventional and organic agriculture (Singh 2014).

8.4.5 Cultural Practices

Cultural practices can control fungal diseases in pulses and other plants. The selection of resistant varieties of plants is necessary and is selected by proper screening in the field. Plantation needs to be done in a well-drained area, with full sunlight. Airflow and ventilation discourage fungal growth, so crowding of plants should be avoided (Bennett et al. 2012). Diseases such as powdery mildew flourish where the nitrogen rate is higher. It promotes tender leaf formation that causes dense strands that are more susceptible to infection. Thus, organic fertilizers or slow-release formulations are good choices. If the infestations are severe, the removal and destruction of the infected plants are effective. Watering plants in the morning is important as it gives the rest of the day time to dry, so that establishment of fungal disease flourishing in wet conditions could be discouraged. Among the treatments of biofungicide, leaf extracts of neem (*Azadirachta indica*), datura (*Datura stramonium*), and debdaru (*Polyalthia longifolia*) showed excellent performance in controlling disease (Hasan et al. 2014).

8.4.6 Organic Control Agent

Sulfur is highly effective against foliar fungal disease including mildews. So, it can be used at a minimum of 7–14 days interval as a protectant. Garlic naturally consists of high levels of sulfur, which can be added with a few cloves crushed in water, like a homemade spray. It is applied as organic fungicide at the first emergence of pathogens and can be repeated if necessary (Djeugap et al. 2014). However, proper timing is vital for successful control, so it should be made sure to begin at the first sign of the disease. Sulfur can cause damage to other edible varieties such as squash; thus another option is to spray it with a solution of baking soda once in a week. It makes the leaf surface unsuitable for the growth of fungal spores by increasing the pH.

8.5 Conclusion

In different aspects of biological control of pulse diseases caused by fungi, a significant improvement has been made, but this area still needs much more investigations and development for the existing problems to be solved. To have strategies in the future with more effective biological control, it is critical for more research to be carried out. On some aspects, novel formulation development, understanding environmental factors' impact on biocontrol agents, mass production of biocontrol agents, and the use of nanotechnology and biotechnology can be used for improving strategies and biocontrol mechanisms (Howell 2007). Future perspectives of pulse disease control are promising and brighter. It is possible to use biological control as a strategy effective for managing diseases of plants, environmental protection, and yield increase and is a sustainable system for agriculture.

Acknowledgment The work is supported by a Network Project, MLP-1005, sponsored by the Council of Scientific and Industrial Research, Ministry of Science and Technology, Government of India, New Delhi. The authors are also thankful to the Director, CSIR-NEIST, Jorhat, Assam, for providing necessary facilities to carry out the work and DBT-BIF Centre, CSIR-NEIST, Jorhat, for providing the computational facilities.

References

Akem C. Ascochyta blight of chickpea: present status and future priorities. Int J Pest Manag. 1999;54:60–52.

Aref AA, Anderson JD. Vigor determination in soybean seed by multiple criteria. Crop Sci. 1973;13:630–3.

Backman PA, Weaver BD, Morgan GJ. Soybean stem canker: an emerging disease problem. Plant Dis. 1985;69:59–21.

Bailey JA, Jeger MJ. *Colletotrichum* biology, pathology and control. J Agric Sci. 1992;56:88–357.

Bailey JA, Jeger MJ. *Colletotrichum*: Biology, Pathology and Control. 2000; CAB International. Wallingford UK.

Bennett AJ, Bending GD, Chandler D, Hilton S, Mills P. Meeting the demand for crop production: the challenge of yield decline in crops grown in short rotations. Biol Rev. 2012;87:52–71.

Busby PE, Ridout M, Newcombe G. Fungal endophytes: modifiers of plant disease. Plant Mol Biol. 2016;90:645–55.

Caffarra A, Rinaldi M, Emanuele E, Vittorio R, Ilaria P. Modelling the impact of climate change on the interaction between grapevine and its pests and pathogens. Agric Ecosyst Environ. 2012;148:45–52.

Chowdhury MEK, Alam MS, Islam MA, Sikdar B, Hasan MF. Identification of pathogen responsible for angular leaf spot disease of *Momordica charantia* and evaluation of biological control system. Int J Sci Res Publ. 2018;8:50–64.

Davidson JA, Kimber RBE. Integrated disease management in ascochyta blight in pulse crops. Eup J Plant Pathol. 2007;119:99–110.

Dilani D, De S, Pedro W, Ades KP, Hybed KD, Paul WJT. Life styles of *Colletotrichum* species and implications for plant biosecurity. Fungal Biol Rev. 2017;31:155–68.

Djeugap JF, Eko D, Nguefack J, Columbus NT, Fontem DA. Effect of organic amendments and fungicide application on potato late blight, bacterial wilt and yield in Cameroon. Int J Agric Res. 2014;5:12–9.

Farouk S, Belal BE, Hael H. The role of some elicitors on the management of Roumy Ahmar grapevines downy mildew disease and it's related to inducing growth and yield characters. Sci Hortic. 2017;225:646–58.

Gardener BBM, Fravel DR. Biological control of plant pathogens: research, commercialization, and application in the USA. Plant Health Prog. 2002;11:1–18.

Gossen BD, Conner RL, Chang KF, Pasche JS, Mc Laren DL, Henriquez MA, Chatterton S, Hwang SF. Identifying and managing root rot of pulses on the northern great plains. Plant Dis. 2016;100:1965–78.

Hasan MM, Islam MR, Hossain I, Shirin K. Biological control of leaf spot of groundnut. J Biosci Agric Res. 2014;1:66–78.

Heath MC. Evolution of plant resistance and susceptibility to fungal parasites. In: Plant Relationships, Part B (G. C. Carroll and P. Tudzynski, eds.), Springer-Verlag, Berlin Heidelberg, 1997; V:257–276.

Howell CR. Mechanisms employed by *Trichoderma* species in the biological control of plant diseases: the history and evolution of current concepts. Plant Dis. 2007;87:75–143.

Kim JS, Lee J, Lee C-h, Woo SY. Activation of pathogenesis-related genes by the Rhizobacterium, Bacillus sp. JS, which induces systemic resistance in tobacco plants. Plant Pathol J. 2015;31:195–201.

Knogge W. Fungal infection of plants. Plant Cell. 1996;8:1711–22.

Maanen AV, Xu XM. Modelling plant disease epidemics. Eur J Plant Pathol. 2003;109:669–82.

Madden LV, Ellis MV, Lalancette N, Hughes M, Wilson LL. Evaluation of a disease warning system for downy mildew of grapes. Plant Dis. 2007;77:861–6.

Meyer V, Andersen MR, Brakhage AA, Braus GH, Mark CX, Timothy CC, et al. Current challenges of research on filamentous fungi in relation to human welfare and a sustainable bioeconomy: a white paper. Fungal Biol Biotechnol. 2016;6:1–3.

Naqvi SAMH. Diseases of fruits and vegetables. Volume II: diagnosis and management, vol. 2. Dordrecht: Springer Netherlands; 2004. p. 97–112.

Nikmaram N, Dar BN, Roohinejad S, Koubaa M, Barba FJ, Greiner R, et al. Recent advances in γ-aminobutyric acid properties in pulses: an overview. J Sci Food Agric. 2017;97:2681–9.

Oumouloud A, Otmani ME, Rouhou HC, Claver AR, Torres RG, Treves RV, et al. Breeding melon for resistance to *Fusarium* wilt: recent developments. Euphytica. 2013;192:152–69.

Pan Z, Li X, Yang XB, Andrade D, Xue L, Mc Kinney N. Prediction of plant diseases through modelling and monitoring airborne pathogen dispersal. Perspect Agric Vet Sci Nutr Nat Resour. 2010;5:45–90.

Pandel S, Sharma M, Kumari S, Gaur P.M, Chen W, Kaur L, MacLeod W, Basandrai A, Bakr A, Sandhu J.S, Tripathi S.H and Gowda C.L.L. Integrated foliar diseases management of legumes. Austr Plant Pathol. 2009; 40(2):149–156.

Passari AK, Lalsiamthari PC, Zothanpuia, Leo VV, Mishra VK, et al. Biocontrol of Fusarium wilt of *Capsicum annuum* by rhizospheric bacteria isolated from turmeric endowed with plant growth promotion and disease suppression potential. Eur J Plant Pathol. 2017;150:831–46.

Passari AK, Upadhyaya K, Singh G, Abdel-Azeem AM, Thankappan S, et al. Enhancement of disease resistance, growth potential, and photosynthesis in tomato (*Solanum lycopersicum*) by inoculation with an endophytic actinobacterium, *Streptomyces thermocarboxydus* strain BPSAC147. PLoS One. 2019;14(7):e0219014.

Pautasso M, Döring TF, Garbelotto M, Pellis L, Jeger MJ. Impacts of climate change on plant diseases – opinions and trends. Eur J Plant Pathol. 2012;133:295–313.

Raimondo ML, Carlucci A. Characterization and pathogenicity of *Plectosphaerella* spp. collected from basil and parsley in Italy. Phytopathol Mediterr. 2018;57:284–95.

Ramaswamy GR, Sohi HS. Studies on spore germination in *Pestalotia psidii*, the causal organism of guava canker. Indian J Mycol Plant Pathol. 1984;14:289.

Reddy TR, Reddy PN, Reddy RR, Reddy SS. Management of turcicum leaf blight of maize caused by *Exserohilum turcicum* in maize. Int J Sci Res. 2013;3:1–4.

Ringer CE, Grybauskas AP. Infection cycle components and disease progress of grey leaf spot on field corn. Plant Dis. 1995;79:24.

Saikia R, Singh T, Kumar R, Srivastava J, Srivastava AK, Arora DK. Role of salicylic acid in systemic resistance induced by *Pseudomonas fluorescens* against *Fusarium oxysporum*. sp. *ciceri* in chickpea. Microbiol Res. 2003;158:203–13.

Saikia R, Kumar R, Singh T, Srivastava AK, Arora DK, Lee MW. Induction of defense related enzymes and pathogenesis related proteins in *Pseudomonas fluorescens*-treated chickpea in response to infection by *Fusarium oxysporum ciceri*. Mycobiology. 2004a;32:47–52.

Saikia R, Singh K, Arora DK. Suppression of *Fusarium*-wilt and charcoal rot of chickpea by *Pseudomonas aeruginosa*. Indian J Microbiol. 2004b;44:10–4.

Saikia R, Srivastava AK, Singh K, Arora DK. Effect of iron availability on induction systemic resistance to *Fusarium* wilt of chickpea by *Pseudomonas* spp. Mycobiology. 2005;33:35–40.

Saikia R, Kumar R, Arora DK, Gogoi DK, Azad P. *Pseudomonas aeruginosa* inducing rice resistance *Rhizoctonia solani*: production of salicylic acid and peroxidases. Folia Microbiol. 2006;5:375–80.

Saikia J, Sarma R.K, Dhandia R, Yadav A, Gupta V.K, Bharali R, Saikia R. Alleviation of drought stress in pulse crops with ACC deaminase producing rhizobacteria isolated from acidic soil of Northeast India. Scientific Reports. 2018;8:3560.

Singh HB. Management of plant pathogens with microorganisms. Proc Indian Natl Sci Acad. 2014;80:443–54.

Sinha P, Rizvi G, Parashar R. Management of wilt disease of pulses: a review. Int J Pure Appl Biosci. 2018;6:696–708.

Sparks A, Kelly L. Mungbean powdery mildew management with fungicide. 2017;GRDC Communities, Australia. https://communities.grdc.com.au/field-crop-diseases/mungbean-powdery-mildew-fungicide/.

Suterman SH, Saefuddin A, Achmad SA. Epidemiology of Needle Blight on Pinus merkusii seedlings incited by Pestalotia theae. J. Manajemen Hutan Tropika (J. Trop. Forest Manag). 2011;10:1–10.

Than PP, Haryudian P, Sitthisack P, Taylor PWJ, Kevin HD. Chilli anthracnose disease caused by *Colletotrichum* species. J Zhejiang Univ Sci B. 2008;9:764–78.

Trivedi S, Srivastava M, Srivastava AK, Ratan V, Shahid M. Status of root and foliar fungal diseases of pulses at different agro-climatic zones of Uttar Pradesh, India. Int J Curr Microbiol Appl Sci. 2017;6:152–65.

Tivoli B, Baranger A, Sivasithamparam K, and Barbetti MJ. Annual Medicago: From a model crop challenged by a spectrum of necrotrophic pathogens to a model plant to explore the Nature of disease resistance. Ann Botany. 2006;98:1117–1128.

Vittorio R, Tito C, Giosuè S, Riccardo B. A mechanistic model simulating primary infections of downy mildew in grapevine. Ecol Model. 2007;212:480–91.

Xin M, Wang Y, Yao Y, Song N, Hu Z, Qin D, et al. Identification and characterization of wheat long non-protein coding RNAs responsive to powdery mildew infection and heat stress by using microarray analysis and SBS sequencing. BMC Plant Biol. 2011;11:61.

Chapter 9
Soil and Crop Health Management for the Cultivation of Pigeon Pea: An Overview of Management Practices

Christy B. K. Sangma

9.1 Background to the Pigeon Pea Crop

Pigeon pea (*Cajanus cajan* L. Millsp.) is a herbaceous pulse crop, under the Leguminaceae family (Fabaceae), predominantly cultivated in tropical and subtropical climatic areas. The crop ranked fifth among the pulse crops in the world contributing 91% to the world's production. In India, it ranked second next to chickpea (occupying 5.13 million hectares area of total 25 million hectares pulse area, 4.23 million tonnes of 18 million tonnes total pulse production and 824 kg ha^{-1} productivity; Anonymous 2015). India is the largest grower of this crop contributing 66% of total production, with the larger portion of production coming from seven states (Maharashtra, Uttar Pradesh, Madhya Pradesh, Karnataka, Rajasthan, Andhra Pradesh and Gujarat). There are largely four types of pigeon pea varieties available, viz. extra-short-duration varieties (<100 days), short-duration varieties (100–120 days) grown in the north-western region, medium-duration varieties (140–180 days) grown in Central India and South India and long-duration varieties (>200 days) grown in the north-eastern plain zone (Singh et al. 2013a; Singh et al. 2013b). The crop is mostly grown as an annual (var. *flavus*) and as a perennial crop (var. *bicolor*) with the rainfed condition in *Kharif* season. Pigeon pea is a drought-enduring crop having a high source of proteins (21–22.3%), vitamins (traces) and minerals such as calcium, magnesium, potassium, phosphorus, iron and fewer amounts of copper and zinc (Saxena et al. 2002). Its carbohydrate content is around 57.2% and very less fat content (around 1.7%), and the crop is largely consumed as "dal" (Singh et al. 2004).

Pigeon pea is a short-day deep-rooted crop and can proliferate as deep as 1.9 m, which enables the plant to explore moisture from deeper soil layers and can bind the soil and reduce erosion (Singh and Russell 1981). It is a widely spaced crop attaining

C. B. K. Sangma (✉)
ICAR Research Complex for NEH Region, Nagaland Centre, Jharnapani, Nagaland, India

B. P. Singh et al. (eds.), *Management of Fungal Pathogens in Pulses*,
Fungal Biology, https://doi.org/10.1007/978-3-030-35947-8_9

a height of 1.5–4 m, grown mostly in less fertile soils and marginal areas with row spacing of 60 cm apart, and comparatively inefficient when grown as the sole crop, due to various reasons, viz. slow initial growth rate, indeterminate growth habit, poor source-sink relationship, poor harvest index, poor biomass production, etc. (Reddy et al. 2011; Nandhini et al. 2015).

9.2 Biological Nitrogen Fixation in Pigeon Pea Crop

Pigeon pea crop has the specialty of biological nitrogen fixation (BNF) and establishes symbiosis with *Bradyrhizobium* spp. (gram-negative, slow grower) which provides more than 90% of nitrogen (N) requirement for its vegetative growth depending on the conduciveness of the growing environment, variety of crop and type of soil (Nambiar et al. 1988; La Favre and Focht 1983). Pigeon pea is known to be the promiscuous legume, which is the capability of the crop to form nodules and nitrogen fixation in symbiotic association with one or more indigenous strains of *Rhizobium*. But results have shown that *Rhizobium* strains are less competent than *Bradyrhizobium* isolates for N_2 fixation in pigeon pea (Anand and Dogra 1991, 1997). In a given season, the crop can fix approximately 40–90 kg ha^{-1} N, and under most favourable environmental conditions, it can fix up to 200 kg ha^{-1} (Kumar Rao and Dart 1987; Adu-Gyamfi et al. 1997; Anonymous 2010). Mhango et al. (2017) reported that the ability to fix total N differed with cropping systems as well, and it is well understood that under intercropping agricultural system, very low level of N was fixed (15 kg N ha^{-1}) as compared to sole pigeon pea (32 kg N ha^{-1}) crop grown in the field. Other than fixing nitrogen, pigeon pea crop is well-known to add biomass to the soil through leftover crop residues (up to 3.1 t ha^{-1}), and the roots of the plant help in mineralizing phosphorus which will be available to the plants.

Temperature is the main factor bi-directionally affecting the legume-*Bradyrhizobium* symbiosis, viz. (i) restricts the development of microsymbionts and (ii) regulates the growth of the acrosymbiont (Hashem et al. 1998; Kuykendall et al. 2000). At low temperature, the height of pigeon pea was reduced and N_2 fixation was hampered. The most favourable temperature was found to be 20–30 °C. Besides temperature, variations in soil pH also influence the survivability of rhizobia. The optimum pH for the rhizobial population is neutral to slightly acidic, and extreme soil pH, viz. acidity, alkalinity and salinity, severely affects the legume production and survival of *Rhizobium* spp. in soil (Slattery et al. 2004). Salt stress and alkalinity also interrupt nodulation, nitrogenase activity and symbiotic N_2 fixation as a whole (Tejera et al. 2006). Though many studies have been conducted for the effects of salinity on N_2 fixation in various leguminous crops, the physiological mechanisms linked are ambiguous. Likewise, in the same manner, acidity also limits the survival of the rhizobial population and reduced nodulation (Taurian et al. 1998).

Bradyrhizobium, an important member of PGPR (plant growth-promoting rhizobacteria), not only carried out nitrogen fixation but also showed indirect effects like phytohormone production, iron chelation, phosphorus solubilization, HCN production, chitinase production, etc. (Deshwal et al. 2003). *Bradyrhizobium* was also found to have an antagonistic effect on soil-borne pathogens (Deshwal et al. 2003).

9.3 Stressors to Pigeon Pea Production

The productivity of pigeon pea in India is 24.7% lower than the world's average. In general, this low productivity is attributed to major barriers including abiotic and biotic factors limiting the maximum yield potential. The major abiotic stresses affecting the crop are temperature, soil acidity, salinity, etc., whereas biotic stresses include the diseases, viz. wilt, sterility mosaic, *Phytophthora* blight, *Alternaria* blight, etc. The crop is also susceptible to various parasitic nematodes, viz. *Meloidogyne javanica* of *Meloidogyne* spp. (root-knot nematode), *Heterodera cajani* (pigeon pea cyst nematode), *Rotylenchus* spp., *Helicotylenchus* spp., etc. (Sharma and McDonald 1990).

9.3.1 Common Diseases of Pigeon Pea in India

Diseases are the main setback in pigeon pea production. The crop is sensitive to hundreds of diseases caused by fungi, bacteria, viruses, mycoplasma-like organisms and nematodes (Reddy et al. 1993). Among the major diseases ($n = 210$) affecting pigeon pea, fungal pathogens are responsible for around 83 diseases, and bacterial diseases are reported to be only 4, whereas the viral and mycoplasma causes 19 and 104 diseases, respectively. Among the pathogens affecting the crop, 98 nos. of pathogens are reported from India (Nene et al. 1989, 1996). Among these pathogens, only a few can cause severe economic losses. Major diseases of pigeon pea which are common in India are given in Table 9.1. Other than these diseases on the standing crops, infected or contaminated seeds also prove hazardous as they cause pre- and post-emergence losses resulting in reduced germination of seeds and reduction of yield and spoiled the quality of seeds during storage. Some researchers (Jalander and Gachande 2011) reported fungal species, viz. *Fusarium oxysporum*, *Fusarium udum*, *Alternaria alternata*, *Aspergillus flavus*, *A. niger*, etc., on stored seed samples of pigeon pea.

Among these diseases, the fungal disease *Fusarium* wilt caused by *Fusarium udum* Bulter was reported to be the highly devastating soil- and seed-borne disease and widely spread in all pigeon pea-growing areas (with maximum damage in states like Maharashtra, Uttar Pradesh, Madhya Pradesh, Bihar and Tamil Nadu) leading to serious yield losses (Pande et al. 2013). Symptoms of this disease include wilting followed by drying up of the crop under field conditions, which show black lines when the infected plant is cut vertically. According to many researchers (Sarojini 1955; Vishwa et al. 2005; Khadse et al. 2015), wilting in pigeon pea was also due to

Table 9.1 Common diseases of pigeon pea, their causal organism and type of damage to the crop

Sl. No.	Disease	Causal organism	Type of damage	Literature
Fungal				
1.	Seedling or seed rot	*Aspergillus flavus*	Reduces protein content in seeds	Sinha and Prasad (1977)
2.	Stem canker, anthracnose	*Colletotrichum capsici*	36.6% yield loss	Tucker (1927)
3.	*Fusarium* wilt	*Fusarium udum*	30–100% yield loss depending on the growth stage of crop	Reddy et al. (1990); Okiror (2002)
4.	*Neocosmospora* root rot	*Neocosmospora vasinfecta*	72.4% wilting percentage	Khadse et al. (2015)
5.	*Phoma* stem canker	*Phoma cajani*	5–50% mortality in plants at maturity stage	Behera et al. (2017)
6.	*Phytophthora* (stem) blight	*Phytophthora drechsleri* f. sp. *cajani*	26.3–98% yield loss	Kannaiyan et al. (1984)
7.	Dry root rot	*Macrophomina phaseolina*	The disease will infect quickly and cause huge economic losses ranging from 10% to 100%. Disease incidence 9–24%	Smita et al. (2015); Maruti et al. (2017)
8.	*Alternaria* blight	*Alternaria alternata*	Disease incidence 20–80% in any kind of cultivar	Sharma et al. (2012)
9.	Wet root rot	*Rhizoctonia solani*	10–50% yield loss	Singh et al. (2009)
Bacterial				
1.	Bacterial leaf spot and canker	*Xanthomonas axonopodis* pv. cajani	40% of disease incidence	Gaikwad and Kore (1981)
2.	Leaf spot	*Cercospora indica*	Yield losses up to 85% and losses are severe when defoliation occurs before flowering and podding	Reddy et al. (1993)
Viruses/mycoplasma				
1.	Sterility mosaic	Virus	With early infection, 95% yield losses occur	Dahal and Neupane (1991)
2.	Phyllody	Mycoplasma-like organism	NA	NA
3.	Pigeon pea mosaic mottle	Viroid	NA	NA
4.	Rosette	Mycoplasma-like organism	NA	NA
Parasitic nematodes (globally cause 13.2–30% yield losses annually in pigeon pea) (Sasser and Freckman 1987; Saxena and Reddy 1987a; Saxena and Reddy 1987b)				
1.	Root-knot nematode	*Meloidogyne* spp.	Yield losses range from 8% to 35%	Sharma et al. (1993)
2.	Pearly root (cyst nematode)	*Heterodera cajani*	Suppresses plant growth by 28% and reduces grain yield by 24% and yield loss up to 49%	Saxena and Reddy (1987a, b); Reddy (1997)

(continued)

Table 9.1 (continued)

Sl. No.	Disease	Causal organism	Type of damage	Literature
3.	Root rot	*Helicotylenchus* spp.	NA	NA
4.	Lance nematode	*Hoplolaimus* spp.	NA	NA
5.	Dirty root	*Rotylenchus* spp.	14–29% yield losses in pigeon pea	Saxena and Reddy (1987a)
6.	Pigeon pea cyst nematode	*Heterodera cajani*	Suppresses plant growth by 28% and reduces grain yield by 24%. Yield losses over 30%	Saxena and Reddy (1987b)

Neocosmospora vasinfecta (Anamorph, *Acremonium* spp.). Besides wilt, *Phytophthora* blight is another major foliar disease of pigeon pea plant which occurs in the seedling stage as well as in the grown-up plants (Pande et al. 2011). In the seedling stage, the symptoms are similar to the damping-off disease, with water-soaked lesions on leaves and breaking of stems, whereas the cankerous outgrowth or galls developed in the stem of the mature plants. The disease is favoured by the high humidity and mostly appears in the low-lying regions of the field and water paths.

9.3.2 Abiotic Stresses Affecting Crop Health

Moisture stress (waterlogging or drought), temperature stress (cold or heat), acidity, alkalinity, salinity, nutrient deficiencies and toxicities, photoperiod, etc. are some of the abiotic factors which affect the production of pigeon pea. Among these stresses, moisture stress is common because pigeon pea is mainly cultivated as a rainfed crop. These abiotic stresses contribute 30–100% of yield losses in the pigeon pea crop (Shabala et al. 1998; Sultana 2010; Choudhary 2013; Pooniya et al. 2015). These stress conditions not only affect the crop directly but also indirectly change the quality and quantity of the microflora of the rhizosphere, adversely affecting the growth and nodulation in the plant. The possibility of the damage and the sensitivity to the diseases, e.g. disease caused by *Macrophomina phaseolina*, also increase under stress conditions.

9.3.2.1 Drought

Although pigeon pea is considered to be a hardy Kharif legume crop because of its vigorous root system, the crop usually suffers from early, intermittent and terminal drought stress, with reduced in yield of about 50% or more (Choudhary 2013; Pooniya et al. 2015). The crop has the four maturity groups, from which extra early and early types complete their life cycle just after the recession of the monsoon season encountering terminal drought in the reproductive phase only. But the medium-

and long-duration pigeon peas face acute soil moisture deficit during the flowering and pod-filling stages which reduced nodule nitrogenase activity (70–90%), followed by the rate of photosynthesis (50–71%) and root and nodule respiration (31–45%). Such a shortage of soil moisture during crucial developmental stages of the plant, like the flowering and pod development stage, decreases the grain yield significantly (Sharma et al. 2012). Drought stress was also found to decrease the rate of photosynthesis (Kawamitsu et al. 2000) and impairs mitosis and cell elongation with a considerable decrease in the number and size of leaves and overall poor performance of the plant as a whole (Hussain et al. 2008). Small-seeded pigeon pea cultivars were reported to be more drought tolerant than the large-seeded cultivars (Kuhad et al. 1989).

9.3.2.2 Waterlogging

Waterlogging is another major limitation for crop production and the productivity of pigeon pea in India. Waterlogging accounts for 1.1 Mha of pigeon pea crop area out of the total area, causing an annual loss of 25–30% (Sultana 2010) and a yield loss of 80–100% (Shabala et al. 1998). Soil types that contribute easily to waterlogging are Vertisols and alluvial soils, with characteristics of high water holding capacity, surface crusting and formation of subsoil pan. Waterlogging can affect pigeon pea during germination and early and late seedling stages and can decrease the height of the plant and delays flowering in surviving plants, ultimately reducing the pod's formation, the number of seeds per pod or the seed yield as a whole.

Pigeon pea requires well-drained soils and is found to be highly susceptible to waterlogging conditions leading to the sudden death of crop (Choudhary et al. 2011). Roots are highly sensitive to anaerobic conditions. The severity of the plants affected due to waterlogging was found to be lower in the intercropped field than sole-cropped fields. The death due to waterlogging may often be confused with the wilt disease of pigeon pea (no sudden death), which can be differentiated with the easy peeling off of bark and presence of brown patches in the collar region in case of waterlogging. Mature plants were found to be more susceptible to waterlogging than the seedlings.

9.3.2.3 Nutrient Stress (Deficiencies and Toxicities)

Nutrient stress occurs due to imbalance application of chemical fertilizers like nitrogen (N), phosphorus (P) and potassium (K), growing of high-yielding varieties, intensive cropping without addition of secondary and micronutrients, no or less use of organic manures, leaching of nutrients under high rainfall and irrigation, conversion of nutrients to unavailable form in problem soils, use of high-analysis fertilizers, negative (−) interaction of micronutrients with other macro-/micronutrients and soil degradation like soil erosion, soil salinity, soil alkalinity, etc. (Reddy et al. 2011; Junjittakarna et al. 2013). Micronutrient deficiencies or toxicities are other limitations for pulse crop production. Restriction of growth and development because of boron (B) toxicity or deficiency is common in leguminous crops (Poulain and Al Mohammad 1995), and these deficiencies or toxicities are more critical in the case of root nodulation than the overall plant growth (Rahman et al. 1999). These micronutrient deficiencies like iron, molybdenum (Mo) and zinc (Zn) or toxicities (boron (B)) can reduce the yield of legume crops at varying magnitudes (Ali et al. 2002).

9.3.2.4 Temperature Stress

Pigeon pea, being a warm-season pulse, an optimum temperature requirement during germination is 30–35 °C, during vegetative stage is 20–25 °C and during flowering and pod-filling stages 15–18 °C and 35–40 °C at maturity, cannot withstand chilling (<15 °C) and frost (Sultana et al. 2014; Rana et al. 2016). The stress considerably upsets the growth, survival and reproductive capacity of the plant when the temperature is lesser than 5 °C. At the freezing temperature, intracellular water gets converted into ice, which in turn causes shrinkage of cells inside the plant, resulting in wilting and death of plants. Singh et al. (1997) studied the effects of low temperature on floral buds and flower drop in the pigeon pea germplasm and observed that long-duration cultivars are well-adapted to cold situations because of their inherent genetic mechanism to cope with very low temperature during reproductive stages. Choudhary (2007) and Sultana et al. (2014) also observed that low-temperature stress (11.4 °C) reduces the number of buds and flowers in pigeon pea.

9.3.2.5 Soil Salinity/Alkalinity Stress

Soil salinity is another major constraint to pigeon pea in regions where it is predominantly grown (Subbarao et al. 1991). Crops cultivated in salt-affected soils experience high osmotic stress conditions, while in alkali soils, nutritional disorders and poor soil physical condition decrease the productivity of the crop. Pigeon pea is very vulnerable to soil salinity and the threshold limit is <1.3 dS m^{-1}. However, some varieties of pigeon pea endured 6–12 dS m^{-1} and even tolerated 3.5 dS m^{-1} salinity through the adaptive mechanisms of the plant (Tayyab et al. 2016). Saline soils can impair the growth and development of the plant, and these cases are mostly observed in irrigated and dryland agriculture. Salinity was found to prolong 50% flowering stage by 1–2 weeks and also delays the peak flowering stage. It stimulates increased flower shedding, reducing the effective number and weight of the pods (Vadez et al. 2007) finally reducing the seeds (Promila and Kumar 1982). During salt stress, improper flower, pollen grain and embryo formation inhibited proper ovule fertilization. Salinity also is known to obstruct the germination of seeds and decreases nodule numbers, ultimately hindering the plant growth and crop yield of pigeon pea (Singh et al. 2016).

9.3.2.6 Soil Acidity

Acid soils occupy considerable areas in different parts of the world. This type of soil is represented by low productivity and infertile areas owing to the toxicities of aluminium (Al) and manganese (Mn) along with deficiency of nutrients, viz. phosphorus (P), calcium (Ca), magnesium (Mg), etc. The key growth-limiting factor in this type of soils is the excess of Al (Singh and Choudhary 2009). In India, acid soils occupy 49 million hectares (Mha), of which 24 Mha have pH below 5.5 (Mandal 1997). Pulses are highly susceptible to soil acidity, and pH less than 5.5 leads to restricted root growth because of Al, Fe and Mn toxicity. Slightly acidic to slightly alkaline soils containing 50% or more sand particles were found to favour disease incidence in susceptible cultivars, and it is also noted that a higher proportion of sand in soil favours occurrence of wilt disease (Shukla and Gupta 1975).

9.3.2.7 Other Constraints

Other limitations to pigeon pea production include faulty sowing practices and seed rate, absence of irrigation facilities, timely availability of quality seeds and use of chemical fertilizers, pesticides, etc. (Ramakrishna et al. 2000; Reddy 2009; Singh et al. 2013a).

In India, pulse crops are cultivated in different agro-climatic regions with varied soil types, rainfall, thermal regimes, topography, etc. This requires precise production techniques with location-specific crop varieties resistant to biotic and abiotic stresses existing in the area. Even the strains used in biofertilizers, biopesticides or biocontrol agents should originate from areas of corresponding agro-climatic regions to be effective and also equally applicable for production technologies like tillage and seeding devices (Singh et al. 2012).

9.4 Soil and Crop Health Management Practices of Pigeon Pea

India is the leading producer (25% of global production), the consumer (27% of world consumption) and the importer (14%) of pulses in the world. Estimates indicate that the country needs a 4.2% growth rate in pulse production annually to ensure the projected demand of 30 million tonnes by 2030. In 2008–2009, the production of pulses was 14.57 million tonnes (Mt) from an area of 22.09 million hectares (FAO 2016). Since then, the acreage under pulse crops remain stagnated for many years and had failed to surpass the demand. As a result, India is compelled to do heavy imports of pulse every year to meet the pulse demand. This situation is likely to get worse shortly considering the increase in population in the country, decrease in the per capita availability of land, competition from other crops and short of advances in technologies. Considering these facts, the Government of India launched various schemes (National Food Security Mission 2007–2008, Accelerated Pulses Production Programme (A3P), Integrated Schemes of Nutrient and Pest Management Programmes, Price Support Policies, etc.) for the promotion of pulses and to increase its productivity and meet the gap between the demand and the supply. Globally, the Food and Agriculture Organization (FAO) had declared 2016 as the "International Year of Pulses" during the 68 Session of the United Nations General Assembly on December 20, 2013 (FAO 2016). This was declared to create awareness about the dietary benefits of pulse crops, increase and sustain the pulse production and ensure self-sufficiency aiming towards food security and nutrition.

Three possible options are available to increase production in pulses (including pigeon pea) and to meet the demands, and these are (1) soil health management, (2) crop health management and (3) increase in acreage under pulses. In this chapter, the increase in acreage under pulses to increase production will not be touched in detail, as it is beyond the scope of this section.

9.4.1 Soil Health Management of Pigeon Pea Crop

Soil is a complex ecosystem in itself, and functioning processes (viz. nutrient cycling and transformations including mobilization, fixation and mineralization, rate of residue decomposition, soil structure formation, etc.) which are governed largely by soil biota community in the ecosystem are the main drivers in regulating the nutrient supplying capacity or fertility of soils. Soil fertility or health depends not only on elemental constituents of soil but also on the quality and quantity of microbes residing in it. These microorganisms are key component of soil biota community, and they are mainly of two types, i.e. the positive effect type or beneficial (PGPRs) and negative effect type or disease-causing organisms, which affect directly or indirectly the productivity and health of any crop (Kennedy and Papendick 1995; Pankhurst et al. 1996). This is true, as the plant-derived nutrients and growth factors, attractants or even inducers of enzymes for microbial colonization from the soil. So, maintaining the soil health by supplying all the necessary elements in the form of organic or inorganic manures is crucial for the crop to remain healthy and productive.

9.4.1.1 Nutrient Management Practices in Pigeon Pea

The poor yield of pigeon pea crop is mainly attributed to their farming in marginal soils with poor management practices of inadequate and imbalanced nutrient application, no application of organic manures and macro- and micronutrients like phosphorus (P), sulphur (S), zinc (Zn), iron (Fe), etc. Hence, nutrient management is found to exert a great influence not only on growth and yield attributes of crops but also for obtaining sustained productivity of pigeon pea. In pigeon pea, the nutrient requirement (recommended dose of fertilizer (RDF) is 20:40:30 or 20:60:30 kg

NPK ha^{-1} depending on the region) is much lesser than cereals due to symbiotic N_2 fixation. But, P deficiency could reduce pigeon pea yields by over 30% (Chauhan et al. 1992). The yield can be increased by about 70% by P application @20 kg ha^{-1} alone, which can be boosted by rhizobial inoculation as well. The pigeon pea crop was reported to consume 56 kg of nitrogen, 5 kg of phosphorus and 22 kg of potassium to produce 1 tonne of grains (Kanwar and Rego 1983). This indicates that the continued crop production without proper nutrient management practices can remove the huge quantity of nutrients leaving the soil deteriorated in due course of time.

Leguminous crop pigeon pea requires a comparatively higher amount of micronutrients, viz. molybdenum (Mo) and iron (Fe), as they are an integral part of the nitrogenase enzyme and required for N fixation (Choudhary et al. 2014). Besides this, boron (B), zinc (Zn) and sulphur (S) deficiencies are reported to be common in pulse-growing areas (Singh et al. 2013a). To tackle some of these deficiencies, application of gypsum or single superphosphate at sowing was carried out, which supply sulphur up to 20–40 kg/ha, and application of $ZnSO_4$ @25–50 kg/ha once in 2 years also addresses these problems effectively and boosts the crop production (Singh et al. 2013b). A balanced dose of nutrients is also important in increasing the yield of pigeon pea. Application of 25:50:25:20 of N:P_2O_5:K_2O:S in kg ha^{-1} and $ZnSO_4$ @15 kg ha^{-1} with organic manures is found optimum for pigeon pea (Anonymous 2012). An unconventional way of nutrient management is to employ soil test-crop response (STCR)-based targeted precision nutrient management practices for higher crop productivity with economic use of chemical fertilizers (Suri and Choudhary 2013). Meena et al. (2012) suggested that the rate of fertilizer application based on soil test yield is found to be higher as compared to conventional methods. Acute deficiency can also be managed by foliar spray of nutrient solutions, e.g. 2% N at flower initiation coupled with manure and fertilizer application (Sharma et al. 2010). Verma et al. (2004) also reported that Zn application in terms of foliar spray @0.5% $ZnSO_4$ also supplements the nutrient requirement directly, which increases plant height (115.5 cm), pods per plant (185 nos.) and seed yield of the crop (1942 kg ha^{-1}) in comparison to other treatments.

Efficient integrated nutrient management practices, especially nitrogen along with biofertilizers, hold a great promise for maintaining the soil health along with the steady supply of nutrients to the plant. Subba Rami Reddy et al. (2011) found that 50% RDF + *Rhizobium* @200 g/kg seed application as basal dose gives better seed yield of pigeon pea. Inoculation of seeds with a combination of biological fertilizers (viz. *Rhizobium* + *Pseudomonas striata*) considerably improved the accumulation of dry matter, the nodulation and the overall yield of pigeon pea (Patil and Padmani 2007). Economic viability of pigeon pea was proved superior with vermicompost application @5 t ha^{-1} plus RDF (20:50 of N and P_2O_5 kg ha^{-1}), gypsum (100 kg ha^{-1}), $ZnSO_4$ @25 kg ha^{-1} and borax @10 kg ha^{-1} and *Rhizobium* (as seed treatment) in Vertisols of Karnataka (Somashekar et al. 2017). DAP application @20 kg P_2O_5 ha^{-1} along with *Bacillus polymyxa* also increases the yield. Application of 40 kg P_2O_5 ha^{-1} through rock phosphate along with either *B. polymyxa* or *Aspergillus awamori* (P solubilizers) was also found to be effective (Anonymous 2001).

Organic components such as enriched composts, FYM, green manure, soil amendments like biofertilizers, etc. affect soil microbial activity, diversity, biomass, respiration and fertility improving the physicochemical characteristics of soils (Grayston et al. 2004). The organic matter also plays a crucial role in maintaining soil physical conditions. Researchers have shown that pulse crop diseases could be reduced significantly with the addition of organic manures, crop residues and organic amendments. These amendments can also reduce the impact of abiotic stresses especially drought stress, salinity conditions (Mayur and Deshmukh 2003), etc. Mayur and Deshmukh (2003) reported that legume wilt incidence was significantly reduced by incorporating de-oiled mustard cake, groundnut cake and FYM into the soils. Kumar et al. (2014) also showed that inoculation of arbuscular mycorrhizal (AM) fungi imparts tolerance to water stress besides phosphorus nutrition in rainfed regions.

Leguminous crops perform well under neutral pH soil condition, and nodulation significantly reduces under the acidity and salinity/alkalinity soil conditions. Liming of acid soils plays the main role in neutralizing the acidity. Liming with dolomitic limestone of 80.3% relative total neutralizing capacity, with an assumption of 60% base saturation for 30 days, is the best way for correcting soil acidity (Singh et al. 2013a). Throughout this phase, soil moisture content of 60% can be maintained for increasing effectiveness. Other soil amendments that can be utilized for correcting soil acidity are basic slag, paper mill sludge, etc. Band application @1/10 of lime requirement plus required doses of fertilizers annually is also found to be economical, practical and effective than lime requirement based on laboratory tests. Furrow application @2–4 q ha^{-1} (particle size below 80 mesh) before planting a crop is the alternate method of application. 40–100% of yield benefit was observed with liming in furrows alone in leguminous crops like pigeon pea, black gram and cowpea grown in low pH soils. Biochar is another such amendment that can ameliorate soil acidity and can reduce the excess of Al. Besides this, biochar is rich in several nutrients, viz. macronutrients (N, P, K), secondary nutrients (Ca, Mg) and micronutrients (Fe, Mn, Zn and Cu), improve water retention and improve soil conditions (Purakayastha et al. 2013). Biochar is applied in many ways, e.g. broadcasting, banding, spot placement, etc.

In the same way, the soil types with pH more than 8 with exchangeable sodium >12–15% require an appropriate management practice for successful cultivation. In such type of soils, mineral calcium helps regulate ion transport into cells of the plant and inhibits Na^+ absorption in pigeon pea (Subbarao et al. 1990). Amendments used for chemical amelioration of saline/alkaline soils are those containing soluble calcium ion in it like gypsum and phosphogypsum which is readily available and cost-effective or acid-forming amendments, viz. pyrites, sulphuric acid, aluminium sulphate, sulphur, etc. These chemical ameliorants are incorporated followed by leaching. For cultivation of crops, gypsum or phosphogypsum is applied at 15–30 days ahead of sowing @75% of gypsum requirement (GR). According to the crop and available sulphur (S) status in soil, gypsum requirement varies from 100 to 200 kg ha^{-1}. Change in yield from 20% to 30% in pulses can be observed with gypsum application alone in soils deficient in sulphur content.

9.4.1.2 Soil Moisture Conservation Practices

Merely 12% of the area under pulses is irrigated in India (Reddy and Reddy 2010), and the major areas come under the rainfed cultivation system. Therefore, soil moisture is the major constraint for pigeon pea in dryland agriculture. Adoption of suitable cultivation techniques is the pre-requisite for conserving soil moisture for maximizing productivity under moisture stress conditions. In drylands, a deep summer ploughing coupled with levelling is essential for moisture conservation; and similarly, supplementary life-saving irrigation during the post-rainy season would be beneficial for increasing productivity. Chaudhary et al. (2003) suggested that in red lateritic areas, grass, *Gliricidia* or *Lantana* mulch applied @8 t ha^{-1} retained significantly higher soil moisture and thereby enhanced pulse crop yield by 2–3 times compared to no-mulch under rainfed conditions. Fertigation also holds a promise for widely spaced crops like pigeon pea, and through this method, 30–50% more area can be irrigated (Singh et al. 2013a, b).

Foliar application of anti-transpirants in pulses is recommended for low productivity of pulses due to erratic and scanty rainfall and prolonged dry spell during flowering and pod-formation stages. Foliar spray of kaolin (6%) with FYM + dust mulch was reported to have a desirable change in the productivity of the pigeon pea + mungbean intercropping system besides reducing evapotranspiration losses of water, suppression of weeds and conservation of soil moisture (Kumar and Rana 2007).

Cover crop also called the *living mulch* also gains considerable attention because of the many benefits it provides for the main crop. It acts as the cover to the soil reducing the erosion as well as reducing evaporation. It accelerates the infiltration of rainwater, improves organic matter content and reduces high temperatures. Cover crops can also suppress soil-borne pathogens, as well as the annual weeds up to a certain extent, and also increase microbial activity. In widely spaced crops like pigeon pea, the cover crop is also a potential option to grow as an intercrop in between main crops. The thick mat of dead plants and residue also acts as the natural mulch for the crop. Examples of cover crops are clovers, hairy vetch, field peas, alfalfa, etc.

9.4.1.3 Manipulation of Rhizospheric Soil for Fungal Disease Management

Soil amendments with decomposable crop residues and oil cakes have been recognized as the most effective method of changing soil and rhizosphere environment, thereby affecting the quality and quantity of soil microflora and fauna, and have already been reported to reduce nematode infestation, viz. *Heterodera cajani* (Pandey and Singh 1990). The application of nitrogen-rich organic amendments releases allelochemicals in the soil through microbial decay, thereby reducing the soil-borne diseases. It also has the potential to suppress the plant pathogens and enhance plant growth-supporting microbes, thereby improving the health of the soil as well as the crop (Papavizas and Lumsden 1980). Oil cakes of neem, mustard, mahua, coconut, linseed and sesame at different concentrations (0.25%, 0.5%, 1.0% and 2.0%) were tested against radial growth of *Fusarium udum* (wilt of pigeon pea). Neem, mustard and mahua oil cakes were found most effective botanicals in reducing fungal growth. The best growth of pigeon pea plants was recorded with mahua oil cake, but the neem oil cake was most effective in controlling *Fusarium* wilt incidence and germination of sclerotia of *Macrophomina phaseolina* (Dwivedi and Dubey 1986). Devadason and Subramanian (2012) observed that the mycelial growth of *Macrophomina phaseolina* can be subdued by the application of a 10% mahua cake. Neem seed oil (*Azadirachta indica*) is well-known for its antiviral, antibacterial, antiprotozoal, anti-insecticidal and antifungal (Murthy and Sirsi 1958; Singh et al. 1980) properties.

PGPRs (plant growth-promoting rhizobacteria *Rhizobium*, *Bradyrhizobium*, *Bacillus*, *Pseudomonas*, *Azospirillum*, *Azotobacter*, *Enterobacter*, *Arthrobacter*, *Burkholderia*, etc.) can influence several plant development mechanisms, viz. atmospheric nitrogen fixation, chelation of iron through siderophore production and making it accessible to the plant root, solubilization of certain minerals (like phosphorus, zinc, potassium, etc.) increasing the mineral uptake by plants and increase in yield by 10–30% and synthesis of phytohormones like indole acetic acid, abscisic acid, gibberellic acid, cytokinins, ethylene, etc. (Gupta et al. 2012; Kennedy et al. 2004; Patten and Glick 2002; Zahir et al. 2003). It suppresses the phytopathogens and synthesizes antifungal metabolites like antibiotics. Root-nodulating rhizobia are also known to reduce the soil-borne root-infecting fungi. Co-inoculation of *P. aeruginosa* and *Bradyrhizobium* has the potential in curbing the root rot disease (*M. phaseolina*, *R. solani* and *F. solani*) on pulses (Ehteshamul-Haque and Ghaffar 1993; Siddiqui et al. 1998).

In general, the soil microflora increases with the addition of nutrients like nitrogen, phosphorus and potassium. This increase of microflora in the rhizosphere zone plays an important role in the disappearance of pathogenic soil *Fusarium* as they are unable to sporulate well. Colonization of *Fusarium* was also found to be low in the presence of minerals like aluminium (Al), cobalt (Co), molybdenum (Mo) and nickel (Ni) (Sulochana 1952). The addition of the solution of micronutrients boron, manganese and zinc is also reported to develop resistance in the host against *Fusarium udum.* Zinc, on the other hand, inhibits spore germination of pathogens and eliminates pathogens quickly from the soil. Similarly, pre-treatment of seeds in Mn solution provides resistance to the plant against infection, or soil amendments at 100 and 200 ppm of Mn exclude fungal spores in the soil (Subramanian 1956).

The cultural operations, viz. deep summer ploughing, soil solarization and adoption of organic amendments, have been reported to control soil-borne diseases (Pande et al. 2013). Soil solarization is a technique of increasing soil temperature during hot summer days usually by covering or mulching the moist soil with a transparent polythene sheet. The idea behind soil solarization is to increase the temperature (35 °C) of moist soil to a lethal range that destroys the soil-borne pathogens directly and indirectly by destroying the resting structures of the soil borne pathogens. The practice of soil solarization is usually very useful under the organic farming system. The wilt disease (*Fusarium udum*), being the soil-borne pathogen, can also be managed by soil solarization. Mihail and Alcorn (1984), on the other hand, reported that soil solarization alone was not effective for controlling *M. phaseolina* in field soils. So, the combined effect of different neem products with an increase in duration of soil solarization gradually decreases propagules of *M. phaseolina* (Dubey et al. 2009). Lodha (1995) also reported reducing the population of *M. phaseolina* by 25–42% by summer irrigation alone.

Fusarium wilt of pigeon pea generally develops in the low-lying regions and in water paths and proliferates rapidly in high humidity areas. The best possible way to reduce mortality by the disease are by sowing pigeon pea on ridges avoiding the maximum exposure to rains and allowing better drainages. Another potential approach to decrease yield losses is by growing varieties of pigeon pea resistant to wilt. Umesha et al. (2017) also reported that the ridge sowing or planting method gives higher grain yield and helps in overcoming the *Phytophthora* blight during waterlogging and avoids wilt disease along with seed treatment with *Rhizobium* + PSB which is found to be beneficial to get a higher yield.

9.4.2 Plant Health Management of Pigeon Pea

The plants' health is usually determined by its environment. Plant environment is in turn comprised of abiotic and biotic factors, which are major constraints in crop production. These factors must be analysed, and effective steps must be undertaken to harness the maximum achievable yields. Since pigeon pea is the second key crop among the pulses in India, crop health management practices are the priority areas, which can be achieved by various following approaches.

9.4.2.1 Intercropping

Pigeon pea is a wide-spaced crop having a deep root system, and the initial slow rate of growth offers a good scope for intercropping with short-duration crops like green gram, black gram or sesame. Intercropping is one of the potent means of increasing total pulse production and income per unit area. In the intercropping system, intercrop has a lower plant population than its sole crop; thus, higher dose of nutrients may help improve yield. Mixed cropping or intercropping of pigeon pea (1:1 and 1:2) with sorghum (*Sorghum vulgare* Pers.) provides the most effective and practical solution by substantially reducing the incidence of wilt (reduce to 4.3% disease incidence) and *Phytophthora* blight incidence (reduce to 1.2% disease incidence) in pigeon pea which is due to the inhibitory effects of exudates and root secretion of hydrocyanic acid (HCN) of sorghum on pathogen (Singh 2000; Agrawal and Tripathi 2003). Intercropping pigeon pea with other crops can also reduce weeds. Kaur et al. (2015) stated that mixed cropping of pigeon pea with soybean (2:4) can subdue the weed growth resulting in more grain yield by 32% when pigeon pea is grown as the sole crop.

One such example is the intercropping of pigeon pea + green gram/black gram which is also helpful in total pulse production and pigeon pea + sesame for enhancing the production of pulses and oilseed (Kumar and Kushwaha 2018). For successful cultivation of any intercropping, plant geometry, suitable varieties and fertilizer management of component crops become important which may vary with crop combination, varieties and location. Pigeon pea crops are fertilized @20 kg N ha^{-1} for sole whereas for intercropping system @20 kg N + 60 kg P2O5 + 20 kg K2O ha^{-1} (Kumar and Kushwaha 2018). Patil et al. (2008) otherwise suggested that for integrated nutrient management system, 50% RDF + vermicompost @3 t ha^{-1} or FYM @5 t ha^{-1} + biofertilizers was found best for intercropping of pigeon pea with pearl millet.

9.4.2.2 Crop Rotation

The rule of thumb for crop rotation is that the same crop should not be grown multiple times. The continuity of the same crop in the same piece of land helps in building up pathogens, insects, weeds, etc. which reduces the yield of the crop. This calls for higher plant protection measures, viz. herbicides, insecticides, pesticides, etc. involving huge cost. Crop rotation is also called *break crop* as it provides a break in a pest, disease or weed through the removal of suitable host and environment. This cropping system also helps in the conservation of soil moisture and building up of organic matter in soil and improves the physical conditions of the soil. The choice of crops in the rotation should include:

(i) N-demanding and N-fixing crop
(ii) Shallow- and deep-rooted crop
(iii) Large root and small root biomass
(iv) Weed-susceptible and weed-suppressing type
(v) Crops with different pest and disease sensitivity
(vi) Grow catch crops, green manures, etc.

Crop rotation is one of the best ways of suppressing the wilt of pigeon pea. Nevertheless, along with crop rotation, field sanitation and deep summer ploughing play a major role in successfully curving the wilt disease. A crop rotation of 4–5 years was noticed to free the field completely of the wilt pathogen. The duration of rotation can be decreased by eliminating the affected roots. Sorghum, pearl millet, cotton and resistant pigeon pea cultivars are recommended as rotation crops (Singh 2000). Natarajan et al. (1985) studied and recorded the impact of cropping systems on the disease. In continuous cropping of pigeon pea, the incidence was as high as 64–69%. A rotation of sorghum and fallow reduced it to 16–31%, and two cycles of sorghum followed by pigeon pea reduced the incidence to 16%. The root exudates of sorghum had a suppressive effect on the pathogen in the soil, thereby suppressing infection of pigeon pea (Singh 2000). Some researchers (Sikora and Greco 1990) have reported reducing the population of nematodes (e.g. *M. incognita*, *M. javanica*, *H. cajani*, etc.) upon the practice of crop rotation.

Usually, the conventional pigeon pea varieties or landraces are long maturing types so normally intercropped with the early-duration cereals and other pulses. Extra short and short varieties have the potential for inclusion as the sole crop into the rotation as an alternative to rice within the rice-wheat systems, especially during periods of water shortage, price incentives and problems of soil fertility.

9.4.2.3 Biopesticides

Constant application of fungicides harms the environment, as the toxic remains persist in soil polluting the entire surroundings. Fungicides also wipe out both the beneficial and disease-causing types and in certain cases even develop resistant species of the pathogen. It also has the chance of exposure to an applicator, and if the fungicides stay in food chains, it is also a threat to the consumer (Hemanth et al. 2016). Biopesticide is a potential substitute for the use of synthetic pesticides in plant disease management. It is eco-friendly with the goal of sustainable agriculture means to control plant pathogens through the use of indigenous or genetically modified organisms (Taylor et al. 1994).

Biocontrol as a component of integrated disease management (IDM) can also be employed effectively to control the pathogen population in the soil. Some of the well-recognized promising biocontrol agents are *Trichoderma* species, *Gliocladium* spp., *Chaetomium* spp., *Pseudomonas fluorescens* and *Bacillus subtilis.* Biocontrol agents efficiently suppress pathogens by suppressing the inoculums potential of the pathogen (Baker and Drury 1981) in forms of antagonism as competition, antibiosis or exploitation. It provides resistance to the host plant by indirectly altering its microenvironment (Mclaughlin et al. 1990). Several researchers have reported a decrease in the incidence of diseases after inoculation of soils or seed treatment with non-phytopathogenic fungi, bacteria and actinomycetes (Chalutz and Wilson 1990; Mandeel and Baker 1991). Biocontrol agent *Trichoderma viride* present in the rhizosphere soil of pigeon pea was found to be efficient in managing the disease caused by *Aspergillus niger*, *Streptomyces* spp., *Penicillium* spp. and *Bacillus* spp. Bapat and Shah (2000) also reported that the strain of *Bacillus brevis* has biological potential against *Fusarium* wilt in pigeon pea. *Aspergillus niger*, *A. flavus*, *A. terreus*, *Penicillium citrinum*, *Trichoderma harzianum* (suppress mycelia growth by 17.52%), *T. viride* (suppress mycelia growth by 43.13%), *T. virens* (suppress mycelia growth by 31.79%) and *Streptomyces griseus* were also demonstrated as potent antagonists for control of pigeon pea wilt disease (Upadhyay and Rai 1987; Chaudhary et al. 2017). Sharma et al. (2018) had observed that integrated disease management by seed treatment with thiram + carbendazim + *Trichoderma viride* + *Rhizobium* + soil application of *Trichoderma viride*, resulting in higher germination percentage (96.8 and 97.2) of pigeon pea, wilt incidence per cent at 60 DAS (2.97 and 3.15), wilt incidence per cent at 150 DAS (9.68 and 7.65) and seed yield (15.10 and 16.28 q ha^{-1}) at two consecutive years, respectively, was found superior over the rest of the treatments. *T. harzianum* application

@10 and 20 g also control the disease by 42.9% and 61.5%, respectively, and *T. harzianum* @10 g can reduce disease by 30% even at the high level of pathogen density (Prasad et al. 2002).

9.4.2.4 Microbial Consortium

Earlier the concept of disease management or biofertilization is to improve the health of the crop or manage the soil health by applying the single antagonist to suppress a single pathogen or to apply a single biofertilizer to enhance specific nutrient requirement in a single cropping system. This concept is also beneficial for the crops, but it is narrow and sometimes not applicable as the crop may suffer from the series of different diseases or may have multi-nutrient deficiencies at a time. So these constraints were analysed, and various microbial consortia have been developed, which contain different compatible inoculants (whether bacteria or fungi), viz. N fixer, P solubilizer, Zn solubilizer, biocontrol agents, etc., in a single product and are available only for the research purpose at some leading agricultural institutes (e.g. Arka Microbial Consortium of ICAR-IIHR, Bangalore; AAU, Jorhat, Assam; etc.) and are not available commercially. These microbial consortia are those PGPRs and biological control agents which possess the secondary effects and otherwise can be applied as biofertilizers, plant strengtheners and biopesticides. For example, *Rhizobium* sp. earlier is mainly used for promoting the soil and plant health but now also has been recognized in decreasing diseases also. These products are environmentally safe and can be used for organic agriculture systems. Rajasekhar et al. (2016) evaluated *Trichoderma harzianum* (TH), *Pseudomonas fluorescens* (PF), *Rhizobium* (Rh) and *Bacillus subtilis* (BS) at variable combinations for plant disease management of pigeon pea or in the form of consortia and have observed that the plant vigour improvement was noticeable and that all the four combinations (TH + PF + BS + Rh) have shown 86%, TH + BS gives 82% and PF + Rh gives 77% disease reduction.

9.4.2.5 Weed Management

Weeds served as the alternate host to most of the disease-causing pathogens and nematodes and even directly reduce the yield of the crop. Pigeon pea is severely infested by weeds mainly as it is a *Kharif* season crop with a slow initial growth rate and wider spacing. This wide spacing allows the weed growth to come up very fast and smothers the crop, which reduces the yield of the crop by 55–60% (Kandasamy 1999). The reduction in yield can go up to 79.93% if the weeds are allowed to grow till the harvest (Talnikar et al. 2008). So, the initial period during the first 6–8 weeks is a crucial phase, and clean cultivation is recommended during this period. Some major weeds of pigeon pea are *Cyperus rotundus*, *Digera alternifolia*, *Parthenium*, *Ageratum conyzoides*, *Euphorbia hirta*, etc. and some of these weeds known to have an allelopathic effect on pigeon pea (Sukhadia et al. 2000).

Field sanitation, clean seeds, application of organic manures, etc. are some of the weed management practices. Besides these, weed destruction by cutting and removal or hand hoeing, hand pulling, tillage, zero tillage, intercropping, crop rotation, closer spacing and flooding or desiccation and burning, soil sterilization and mulching can be followed to decrease the weed infestation. Mulching is found to be efficient in controlling annual weeds and some perennial weeds like *Cynodon dactylon*, *Sorghum halepense*, etc. (Talnikar et al. 2008). Sugarcane trash mulching @8 t/ha is also effective for control of weeds, increasing yield, conservation of soil moisture and moderation of soil temperature in pigeon pea (Gajera et al. 1998). Chemical weed control (like pendimethalin @1.25 kg ha^{-1} for broad-leaved weed or fluchloralin 0.5–1.0 kg ha^{-1} or oxadiazon 0.75 kg ha^{-1} and quizalofop-p-ethyl @0.5% or alachlor @2 kg ha^{-1} for duration legumes) is also found to be most promising (Kaur et al. 2015).

9.4.2.6 Manipulation in Cultivation Practices

Sowing of pigeon pea by broadcasting on flatbed is the traditional method of pigeon pea cultivation which produces low yield and is at the same time prone to waterlogging conditions. This problem can be tackled by sowing crops on raised broad bed furrow, which drains out excess water easily, also saves irrigation water (16–20%) and induces less crop lodging. Ridge and furrow systems of planting are usually beneficial when saline irrigation waters are used. This method is also successful in draining excess water from crop root zone, reduces the incidence of insect pests and diseases and results in 25–30% higher yield in *Kharif* pulses over flatbed planting (Das et al. 2014). Tillage is necessary for obtaining ideal conditions for proper seed germination, seedling establishment and growth of crops. For pulses, deep ploughing results in better moisture conservation and better root proliferation. Deep ploughing in summer and exposing the soil to the sun effectively reduce *Fusarium* wilt and root rot in chickpea and pigeon pea. Another option is zero-tillage practices, which minimize the soil erosion, and conservation tillage system which conserves soil moisture in moisture-deficit areas (Das et al. 2014). Apart from this cultivation practices, plant diseases can be kept under control by adopting good field sanitation by removing the infected plants and their debris which keeps the primary inoculum at a low level. Practices like timely sowing of the crop, proper spacing, proper depth of sowing, etc. are also helpful in reducing the diseases.

9.4.2.7 Resistant Varieties

Selection of suitable varieties or cultivars of pigeon pea to different regions and weather conditions, tolerant or resistant varieties to abiotic and biotic stresses, etc. is an important option to improve plant growth, disease management and productivity of pigeon pea in any condition as plant response to abiotic and biotic stresses is found to be variety- or cultivar-specific (Maheswari et al. 2015). In drought and heat stress areas with low rainfall and terminal drought conditions, early maturing varieties (short-duration crops) are widely used.

Acknowledgements The author is thankful to ICAR Research Complex for NEH Region, Nagaland Centre, Jharnapani, Medziphema, Nagaland 797 106 for the facilities and the financial support.

References

Adu-Gyamfi JJ, Ito O, Yoneyama T, Devi TG. Nitrogen management and biological nitrogen fixation in sorghum/pigeon pea intercropping on Alfisols of the semi-arid tropics. Soil Sci Plant Nutr. 1997;43:1061–6.

Ali MEK, Inanaga S, Sugimoto Y. Sources of resistance to *Fusarium* wilt of chickpea in Sudan. Phytopathol Mediterr. 2002;41:163–9.

Agarwal SC, Tripathi AK. Efficacy of different intercropping systems on wilt and Phytophthora blight incidence and yield of pigeon pea. Indian J Pulses Res. 2003;16(1):71–72.

Anand RC, Dogra BC. Physiological and biochemical characteristics of fast and slow growing *Rhizobium* sp., from pigeon pea (*Cajanus cajan*). J Appl Microbiol. 1991;70:197–202.

Anand RC, Dogra RC. Comparative efficiency of *Rhizobium/Bradyrhizobium* sp. strains in nodulating *Cajanus cajan* in relation to characteristic metabolic enzyme activities. Biol Fertil Soils. 1997;24:283–7.

Anonymous. Punjab State Council for Science and Technology (A State Government Undertaking). Chandigarh (India). 2001.

Anonymous. Annual report of AICRP on pigeon pea. Kanpur: Indian Institute of Pulses Research; 2010.

Anonymous. Status of pigeon pea research in Karnataka. Bangalore: Published by Principal Scientist AICRP on Pigeon pea UAS, GKVK; 2012.

Anonymous. Agriculture statistics at glance. Ministry of agriculture and farmer welfare. New Delhi: Government of India; 2015.

Baker R, Drury R. Inoculum potential and soilborne pathogens: the essence of every model is within the frame. Phytopathology. 1981;71:363.

Bapat S, Shah AK. Biological control of fusarial wilt of pigeon pea by *Bacillus brevis*. Can J Microbiol. 2000;46:125–32.

Behera S, Rai AK, Rout R. Factor affecting growth of *Phoma cajani* causing stem canker in pigeonpea. Int J Curr Microbiol App Sci. 2017;6:2991–5.

Chalutz E, Wilson CL. Post harvest biocontrol of green and blue mold and sour rot of citrus fruit by *Debaryomyces hansenii*. Plant Dis. 1990;74:134–7.

Chaudhary B, Kumar S, Kushwaha SK. Bio-efficacy of *Trichoderma* species against pigeon pea wilt pathogen. J Appl Nat Sci. 2017;9:2327–31.

Chauhan YS, Saxena NP, Johansen C. Abiotic factors limiting chickpea and pigeon pea production. In: Sachan JN, editor. Proceedings of national symposium on new frontiers in pulses research and development. Kanpur: Directorate of Pulses Research; 1992. p. 111–23.

Choudhary AK. Technological and extension yield gaps in pulses in Mandi district of Himachal Pradesh. Indian J Soil Conserv. 2013;41:88–97.

Choudhary AK, Sultana R, Pratap A, Nadarajan N, Jha UC. Breeding for abiotic stresses in pigeon pea. J Food Leg. 2011;24:165–74.

Choudhary AK, Pooniya V, Bana RS, Kumar A, Singh U. Mitigating pulse productivity constraints through phosphorus fertilization–a review. Agric Rev. 2014;35:314–9.

Choudhary AK. Selection criteria for low temperature tolerance in long-duration pigeonpea. In: Abstract published in the national symposium on legumes for ecological sustainability: emerging challenges and opportunity. Indian Institute of Pulses Research, Kanpur, 2007: pp. 266.

Chaudhary RS, Patnaik US, Dass A. Efficacy of mulches in conserving monsoon of moisture for rabi crops. J Indian Soc Soil Sci. 2003;51(4):495–498.

Dahal G, Neupane KR. Incidence and effect of sterility mosaic disease of pigeon pea in Nepal. Int Pigeon Newsl. 1991;13:23–4.

Das SK, Avasthe RK, Singh R, Babu S. Biochar as carbon negative in carbon credit under changing climate. Curr Sci. 2014;107:1090–1.

Deshwal VK, Pandey P, Kang SC, Maheshwari DK. Rhizobia as a biocontrol agent against soil borne plant pathogenic fungi. Indian J Exp Biol. 2003;41:1160–4.

Devadason A, Subramanian S. Effects of biocontrol agents and plant products on *Macrophomina phaseolina* and colchicine content in *Gloriosa superba*. Plant Prot Sci. 2012;48:110–5.

Dubey RC, Kunar H, Pandey RR. Fungitoxic effect of neem extracts on growth and sclerotial survival of *Macrophomina phaseolina* in vitro. J Am Sci. 2009;5(5):17–24.

Dwivedi RS, Dubey RC. Effect of volatile and non-volatile fraction of two medicinal plants on germination of *Macrophomina phaseolina*. Trans Br Mycol Soc. 1986;87:326–8.

Ehteshamul-Haque S, Ghaffar A. Use of rhizobia in the control of root rot diseases of sunflower, okra, soybean and mungbean. J Phytopathol. 1993;138:157–63.

Food and Agriculture Organization. International Year of Pulses. 2016 https://en.wikipedia.org/wiki/International_Year_of_Pulses

Gaikwad BM, Kore SS. Bacterial leaf spot and stem canker of pigeon pea (*Cajanus cajan*) caused by *Xanthomonas cajani*. Indian J Mycol Plant Path. 1981;11:50–6.

Gajera MS, Ahlawat RPS, Ardeshna RB. Effect of irrigation schedule, tillage depth and mulch on growth and yield of winter pigeon pea (*Cajanus cajan*). Indian J Agron. 1998;43:689–93.

Grayston SJ, Campbell CD, Bardgett RD, Mawdsley JL, Clegg CD, Ritz K, et al. Assessing shifts in soil microbial community structure across a range of grasslands of differing management intensity using CLPP, PLFA and community DNA techniques. Appl Soil Ecol. 2004;25:63–84.

Gupta V, Llewellyn R, McBeath T, Kroker S, Davoren B, McKay A, et al. Break crops for disease and nutrient management in intensive cereal management. In: Australian Agronomy Conference, Australian Society of Agronomy, Armidale, NSW. 2012. http://www.regional.org.au/au/asa/2012/nutrition/7961_vadakattugupta.htm.2012.

Hashem FM, Swelim DM, Kuykendall LD, Mohamed AI, Abdel-Wahab SM, Hegazi NI. Identification and characterization of salt- and thermotolerant *Leucaena*-nodulating *Rhizobium* strains. Biol Fertil Soils. 1998;27:335–41.

Hemanth G, Kumar PKR, Niharika PS, Kolli SK. Fungicides effect on soil micro flora in Tekkali Mandal, Srikakulam (Dist.). Int J Res Dev Pharm Life Sci. 2016;5:2245–50.

Hussain AI, Anwar F, Sherazi STH, Przybylski R. Chemical composition. Antioxidant and antimicrobial activities of basil (*Ocimum basilicum*) essential oils depends on seasonal variations. Food Chem. 2008;108:986–95.

Jalander V, Gachande BD. Seed borne mycoflora of different varieties of pigeon pea (Cajanus cajan (L.) Millsp.). Bioinfolet. 2011;8:167–168.

Junjittakarna J, Pimratch S, Jogloy S, Htoon W, Singkham N, Vorasoot N, Toomsan B, Holbrook CC, Patanothai A. Nutrient uptake of peanut genotypes under different water regimes. Int J Plant Prod. 2013;7:1735–6814.

Kandasamy OS. Effect of herbicide with and without manual weeding on weed and yield of rainfed pigeonpea (*Cajanus cajan*). Leg Res. 1999;22:172–6.

Kannaiyan J, Nene YL, Reddy MV, Ryan JG, Raju TN. Prevalence of pigeon pea diseases and associated crop losses in Asia, Africa and the Americas. Trop Pest Manag. 1984;30:62–71.

Kanwar JS, Rego TJ. Fertilizer use and watershed management in rainfed areas for increasing crop production. Fertilizer News. 1983.

Kaur R, Raj R, Das TK, Shekhawat K, Singh R, Choudhary AK. Weed management in pigeonpea-based cropping systems. Indian J Weed Sci. 2015;47:267–76.

Kawamitsu Y, Driscoll T, Boyer JS. Photosynthesis during desiccation in an intertidal alga and a land plant. Plant Cell Physiol. 2000;41:344–53.

Kennedy AC, Papendick RI. Microbial characteristics of soil quality. J Soil Water Conserv. 1995;50:243–8.

Kennedy IR, Choudhury ATMA, Kecskes ML. Non-symbiotic bacterial diazotrophs in crop-farming systems: can their potential for plant growth promotion be better exploited? Soil Biol Biochem. 2004;36:1229–44.

Khadse RR, Giri GK, Raut S, Bhoye BB. In vitro efficacy of fungicides and bioagents against wilt of pigeon pea caused by *Neocosmospora vasinfecta*. Sci Int. 2015;3(3):82–4. https://doi.org/10.17311/sciintl.

Kuhad MS, Nandwal AS, Kundu BS. Physiological responses of pigeon pea (*Cajanus cajan* L.) genotypes to water stress. Indian J Plant Physiol. 1989;32:212–6.

Kumar U, Kushwaha HS. Studies on nutrient management in pigeon pea (*Cajanus cajan* (L) Millsp) based intercropping system of urd bean, sesame and mung bean. J Pharmacogn Phytochem. 2018;7:490–4.

Kumar ATS, Rana KS. Effect of cropping systems, moisture conservation practices and fertility levels on growth and yield of pigeon pea and mungbean in intercropping system. Ann Agric Res New Ser. 2007;28:63–7.

Kumar Rao JVDK, Dart PJ. Nodulation, nitrogen fixation and nitrogen uptake in pigeon pea (*Cajanus cajan* (L.) Millsp) of different maturity groups. Plant Soil. 1987;99:255.

Kumar A, Suri VK, Choudhary AK. Influence of inorganic phosphorus, VAM fungi and irrigation regimes on crop productivity and phosphorus transformations in okra (*Abelmoschus esculentus* L.)–pea (*Pisum sativum* L.) cropping system in an acid Alfisol. Commun Soil Sci Plant Anal. 2014;45:953–67.

Kuykendall LD, Hashem FM, Dadson RB, Elkan GH. Nitrogen fixation. In: Lederberg J, editor. Encyclopedia of microbiology, vol. 3. 2nd ed. San Diego, CA: Academic Press; 2000. p. 392–406.

La Favre JS, Focht DD. Comparison of N_2 fixation and yields in *Cajanus cajan* between hydrogenase-positive and hydrogenase-negative rhizobia by in situ acetylene reduction assays and direct 15N partitioning. Plant Physiol. 1983;72:971–7.

Lodha S. Soil solarization, summer irrigation and amendments for the control of *Fusarium oxysporum* f. sp. cumini and *Macrophomina phaseolina* in arid soils. Crop Prot. 1995;14:215–9.

Mandal SC. Introduction and historical overview, in acid soils of India. In: Mahapatra IC, Mandal SC, Misra C, Mitra GN, Panda N. (Eds). ICAR, New Delhi, India, 1997: pp. 3–24.

Maheswari M, Sarkar B, Vanaja M, Srinivasarao M, Srinivasarao Ch, Venkateswarlu B, Sikka AK. Climate resilient crop varieties for sustainable food production under aberrant weather conditions. Technical Bulletin, ICAR-Central Research Institute for Dryland Agriculture, Hyderabad. 2015. p. 47.

Mandeel Q, Baker R. Mechanisms involved in biological control of *Fusarium* wilt of cucumber with strains of non-pathogenic *Fusarium oxysporum*. Phytopathology. 1991;81:462–9.

Maruti SAS, Sunkad G, Mahalinga D, Patil MG. Incidence of dry root rot of pigeon pea in north eastern Karnataka, India. Int J Curr Microbiol App Sci. 2017;6:1071–8.

Mayur D, Deshmukh VV. Effect of bio-agents and soil amendments on chickpea wilt caused by *Fusarium oxysporum*. Res Crops. 2003;4:141–3.

Mclaughlin MJ, Malik KA, Memon KS, Adris M. The role of phosphorus in N fixation in upland crops. In: Phosphorus requirement for sustainable agriculture in Asia and Oceania. S.I.: IRRI; 1990.

Meena MC, Dwivedi BS, Singh D, Sharma BM, Kumar K, Rana DS. Effect of integrated nutrient management on productivity and soil health in pigeon pea (*Cajanus cajan*)-wheat (*Triticum aestivum*) cropping system. Indian J Agron. 2012;57:333–7.

Mhango WG, Snapp S, Kanyama Phiri Y. Biological nitrogen fixation and yield of pigeon peas and groundnut: quantifying response on smallholder farms in northern Malawi. Afr J Agric Res. 2017;12:1385–94.

Mihail JD, Alcorn SM. Effect of soil solarisation on *Macrophomina phaseolina* and *Sclerotium rolfsii*. Plant Dis. 1984;68:156–9.
Murthy SP, Sirsi M. Pharmacological studies on *Melia azadirachta*: Part I. Antibacterial, antifungal and antitubercular activity of neem oil and its fractions. Symp Utilization Indian Med PL, Lucknow. 1958. p. 55.
Nambiar PTC, Rego TJ, Rao BS. Nitrate concentration and nitrate reductase activity in the leaves of three legumes and three cereals. Ann Appl Biol. 1988;112:547–53.
Nandhini DU, Vimalendran L, Latha KR, Sangamithra S, Kalaiyarasan V. A review on biological advantages of pigeonpea intercropping influenced by different cropping geometries. Int J Agric Sci Res. 2015;5:103–12.
Natarajan M, Kannaiyan J, Willey RW, Nene YL. Studies on effects of cropping system on *Fusarium* wilt of pigeon pea. Field Crop Res. 1985;10:333–46.
Nene YL, Sheila VK, Sharma SB. A world list of chickpea (Cicer arietinum L.) and pigeon pea (Cajanus cajan (L.) Millsp.) pathogens. Legumes pathology progress report-7. Patancheru 502 324, Andhra Pradesh, India: International Crops Research Institute for the Semi-Arid Tropics, 1989;7:1–23.
Nene YL, Sheila VK, Sharma SB. A world list of chickpea and pigeon pea pathogens. 5th ed. Patancheru: International Crops Research Institute for the Semi-Arid Tropics; 1996.
Okiror MA. Genetics of resistance to *Fusarium udum* in pigeon pea (*Cajanus cajan* (L.) Millsp.). Indian J Genet Plant Breed. 2002;62:218–20.
Pande S, Sharma M, Naga Mangla U, Ghosh R, Sundaresan G. Phytophthora blight of pigeon pea (*Cajanus cajan* (L.) Millsp.): an updating review of biology, pathogenicity and disease management. Crop Prot. 2011;30:951–7.
Pande S, Sharma M, Guvvala G. An updated review of biology, pathogenicity, epidemiology and management of wilt disease of pigeon pea (*Cajanus cajan* (L.) Millsp.). J Food Leg. 2013;26:1–14.
Pandey G, Singh KP. Effect of organic amendments on soil microflora and nematode fauna with special reference to Meloidogyne incognita in chick pea. New Agric. 1990;1:65–70.
Pankhurst CE, Ophel-Keller K, Doube BM, Gupta VVSR. Biodiversity of soil microbial communities in agricultural systems. Biodivers Conserv. 1996;5:197–209.
Papavizas GC, Lumsden RD. Biological control of soil borne fungal propagules. Annu Rev Phytopathol. 1980;18:389–413.
Patil AB, Padmani DR. Effect of integrated nutrient management on growth and yield of pigeon pea (*Cajanus cajan* L. Millsp.). Int J Agric Sci. 2007;3:49–51.
Patil HM, Tuwar SS, Wani AG. Studies on integrated nutrient management for pigeonpea + pearlmillet intercropping system under dryland conditions. Int J Agric Sci. 2008;4:335–9.
Patten CL, Glick BR. Role of *Pseudomonas putida* indole acetic acid in development of the host plant root system. Appl Environ Microbiol. 2002;68:3795–801.
Pooniya V, Choudhary AK, Dass A, Bana RS, Rana KS, Rana DS, et al. Improved crop management practices for sustainable pulse production: an Indian perspective. Indian J Agric Sci. 2015;85:747–58.
Poulain D, Al Mohammad H. Effects of boron deficiency and toxicity on faba bean (*Vicia faba* L.). Eur J Agron. 1995;4:127–34.
Prasad RD, Rangeshwaran R, Hegde SV, Anuroop CP. Effect of soil and seed application of *Trichoderma harzianum* on pigeon pea wilt caused by *Fusarium udum* under field conditions. Crop Prot. 2002;21:293–7.
Promila K, Kumar S. Effect of salinity on flowering and yield characters in pigeon pea. Indian J Plant Physiol. 1982;25:252–7.
Purakayastha TJ, Pathak H, Savita K. Stability of biochar carbon – its implication on carbon sequestration and microbial activities in soil. In: Proc. 100th Indian science congress, part II, abstracts of oral presentation. University of Calcutta, Kolkata. 2013. pp. 287–288.
Rahman MH, Kawai S, Alam S, Hoque S, Tanaka A, Ito M. Effect of soil compaction on plant growth in an Andisol. Jpn J Trop Agric. 1999;4:129–35.
Rajasekhar L, Sain SK, Divya J. Evaluation of microbial consortium for plant health management of pigeon pea. Int J Plant Anim Environ Sci. 2016;6:107–13.

Ramakrishna A, Gowda CLL, Johansen C. Management factors affecting legumes production in the Indo-Gangetic plain. In: Johansen C, Duxbury JM, Virmani SM, Gowda CLL, editors. Legumes in rice and wheat cropping systems of the Indo-Gangetic plain-constraints and opportunities. Patancheru: ICRISAT; 2000. p. 156–65.

Rana DS, Dass A, Rajanna GA, Kaur R. Biotic and abiotic stress management in pulses. Indian J Agron. 2016;61:238–48.

Reddy AA. Pulses production technology: status and way forward, review of agriculture, economic & political weekly. Rev Agric. 2009;44(52):73–80.

Reddy AA, Reddy GP. Supply side constrains in production of pulses in India: a case study of lentil. Agric Econ Res Rev. 2010;23:129–36.

Reddy MV, Sharma SB, Nene YL. Pigeon pea: disease management. In: Nene YL, Hall SD, Sheila VK, editors. The pigeon pea. Wallingford: CAB International; 1990. p. 303–48.

Reddy MV, Raju TN, Sharma SB, Nene YL, McDonald D. Handbook of pigeon pea diseases. Information Bulletin no. 42. Patancheru: International Crops Research Institute for the Semi-Arid Tropics; 1993. p. 64.

Reddy ASR, Babu JS, Reddy MCS, Khan MM, Rao MM. Integrated nutrient management in pigeon pea (*Cajanus cajan* L.). Int J Appl Biol Pharm Technol. 2011;2:462–70.

Reddy BMR. Status of nematode problems an d research in Karnataka. In: Sharma SB (Ed.) Diagnosis of key nematode pests of chickpea and pigeon pea and their management: Proceedings of a Regional Training Course. ICRISAT, Patancheru, India Patancheru, Andhra Pradesh, India: International Crops Research Institute for the Semi-Arid Tropics, 1997: pp. 87–91.

Sarojini TS. Soil conditions and root diseases IX *Neocosmospora vasinfecta* Smith disease of *Cajanus cajan*. Rev Appl Mycol. 1955;34:727.

Sasser JN, Freckman DW. A world perspective on nematology: the role of the society. In: Veech JA, Dickson DW, editors. Vistas on nematology. Hyattsville: Society of Nematologists; 1987. p. 7–14.

Saxena R, Reddy DDR. Crop losses in pigeon pea and mungbean by mungbean cyst nematode, *Meloidogyne cajani*. Indian J Nematol. 1987a;17:91–4.

Saxena R, Reddy DDR. Crop losses in pigeon pea and mungbean by pigeon pea cyst nematode, *Heterodera cajani*. Indian J Nematol. 1987b;17:91–4.

Saxena KB, Kumar RV, Rao PV. Pigeon pea nutrition and its improvement. J Crop Prod. 2002;5:227–60.

Shabala S, Newman I, Whittington J, Juswono U. Protoplast ion fluxes: their measurement and variation with time, position and osmoticum. Planta. 1998;204:146–52.

Sharma SB, McDonald D. Global status of nematode problems of groundnut, pigeonpea, chickpea, sorghum, and pearl millet and suggestions for future work. Crop Prot. 1990;9:453–8.

Sharma SB, Nene YL, Reddy MV, McDonald D. Effect of *Heterodera cajani* on biomass and grain yield of pigeon pea on Vertisol in pot and field experiments. Plant Pathol. 1993;42:163–7.

Sharma P, Shukla MK, Sammis TW, Adhikari P. Nitrate-nitrogen leaching from onion bed under furrow and drip irrigation systems. Appl Environ Soil Sci. 2012;2012:1–17. https://doi.org/10.1155/2012/650206.

Sharma RL, Mishra T, Bhagat R, Swarnkar V, Kumar K. Integrated disease management for pigeon pea wilt caused by *Fusarium udum*. Int J Chem Stud. 2018;6:2748–51.

Sharma A, Nakul HT, Jelgeri BR, Surwenshi A. Effect of micronutrients on growth, yield and yield components in pigeon pea (Cajanus cajan L. Millsp.). Res J Agric Sci. 2010;1(2):142–144.

Shukla UC, Gupta BL. Response to Mn application and evaluation of chemical extractants to determine available Mn in some arid brown soils of Haryana. J Indian Soc Soil Sci. 1975;23:357–60.

Siddiqui ZA, Mahmood I, Hayat S. Biocontrol of *Heterodera cajani* and *Fusarium udum* on pigeon pea using *Glomus mosseae*, *Paecilomyces lilacinus* and *Pseudomonas fluorescens*. Thai J Agric Sci. 1998;31:310–21.

Sikora RA, Greco N. Nematode parasites of food legumes. In: Luc M, Sikora RA, Bridge J, editors. Plant parasitic nematodes in subtropical and tropical agriculture. Wallingford: CAB International; 1990. p. 181–235.

Singh VP. Planting geometry in maize (*Zea mays*) and blackgram (*Phaseolus mungo*) intercropping system under rainfed low hill valley of Kumaon. Indian J Agron. 2000;45:274–8.

Singh D, Choudhary AK. Screening of pigeonpea genotypes for tolerance to aluminium toxicity. Paper presented in the International conference on Grain Legumes: Quality Improvement, Value Addition and Trade organized by Indian Society of Pulses Research and Development at IIPR, Kanpur from Feb 14–16, 2009.

Singh SB, Chhabra R, Abrol IP. Effect of soil sodicity on the yield and chemical composition of cowpea (*Vigna sinensis* L.) grown for fodder. Indian J Agric Sci. 1980;50:852–6.

Singh Y, Gaur NS, Singh D, Singh D. Response of pigeon pea (*Cajanus cajan*) to NPK in western part of Uttar Pradesh. Ann Agric Res. 1997;15:495.

Singh V, Pandey PC, Jain DK. A text book of botany angiosperms. Meerut: Rastogi Publication; 2004.

Singh MV, Narwal RP, Bhupal RG, Patel KP, Sadana US. Changing scenario of micronutrient deficiencies in India during four decades and its impact on crop responses and nutritional health of human and animals. In: The proceedings of the international plant nutrition colloquium XVI. Department of Plant Sciences, UC Davis. 2009.

Singh NK, Gupta DK, Jayaswal PK, Mahato AK, Dutta S, Singh S, et al. The first draft of the pigeon pea genome sequence. J. Plant Biochem Biotechnol. 2012;21:98. https://doi.org/10.1007/s13562-011-0088-8.

Singh AK, Meena MK, Bharati R, Gade RM. Effect of sulphur and zinc management on yield, nutrient uptake, changes in soil fertility and economics in rice (*Oryza sativa*) – lentil (*Lens culinaris*) cropping system. Indian J Agric Sci. 2013a;83:344–8.

Singh AK, Rai VP, Chand R, Singh RP, Singh MN. Genetic diversity studies and identification of SSR markers associated with *Fusarium* wilt (*Fusarium udum*) resistance in cultivated pigeon pea (*Cajanus cajan*). J Genet. 2013b;92:273–80.

Singh S, Grover P, Kaur J, Singh I, Kaur J, Singh P, et al. Genetic variability of pigeon pea (*Cajanus cajan* (L.) Millsp.) for Waterlogging and salinity tolerance under in vitro and in vivo conditions. Am J Exp Agric. 2016;12:1–13.

Singh S, Russell MB. Water use by a maize/pigeon pea intercrop on a deep Vertisol. In: Nene YL, Kumble V (Eds.) Proceedings of the International Workshop on Pigeon peas, ICRISAT (International Crops Research Institute for the Semi-Arid Tropics), Volume 1, Patancheru, A.P., India, 1981.

Sinha MK, Prasad T. Deterioration of arhar seeds by Aspergillus flavus. Indian phytopathol. 1977;30(1):70–72.

Smita KP, Rajeswari E, Alice D, Raguchander T. Assesment of vascular wilt and dry root rot of pigeon pea in Tamil Nadu. International J of Tropical Agri. 2015;33(3):2145–2151.

Slattery JF, Pearce DJ, Slattery WJ. Effects of resident communities and soil type on the effective nodulation of pulse legumes. Soil Biol Biochem. 2004;36:1339–46.

Somashekar TN, Tiwari A, Wani SP, Tedia K. Economics of pigeon pea as influenced by method of planting and integrated nutrient management in pigeon pea. Int J Chem Stud. 2017;5:1404–6.

Subba Rami Reddy A, Sateesh Babu J, Chandra Sekhar Reddy M, Mujeeb Khan MD, Murali Rao M. Integrated nutrient management in pigeon pea (*Cajanus cajan*). Int J Appl Bio Pharm Technol. 2011;2:467–70.

Subbarao GV, Johansen C, Rao JVDKK, Jana MK. Salinity tolerance in F1 hybrids of pigeon pea and a tolerant wild relative. Crop Sci. 1990;30:785–8.

Subbarao GV, Johansen C, Jana MK, Rao JVDKK. Comparative salinity responses among pigeon pea genotypes and their wild relatives. Crop Sci. 1991;31:415–8.

Subramanian D. Studies on the control of fungal wilts of plants. Thesis submitted for the Degree of Doctor of Philosophy. University of Madras. 1956.

Sultana N. Nutrition content and heavy metal contamination in some roadside soils and grasses of Dhaka City, Bangladesh. M. Sc. Thesis, Department of Agricultural Chemistry, Bangladesh Agricultural University, Mymensingh, 2010.

Sukhadia NM, Ramani BB, Modhwadia MM, Asodaria KB. Integrated weed management in pigeon pea (*Cajanus cajan* L. Millsp.). Gujarat Agric Univ Res J. 2000;25:1–4.

Sulochana CB. Soil conditions and root diseases VI. Germination of conidia of *Fusarium vasinfectum* in micro-element amended soils. Proc Indian Acad Sci Sec B. 1952;36:229–33.

Sultana R, Choudhary AK, Pal AK, Saxena KB, Prasad BD, Singh RG. Abiotic stresses in major pulses: current status and strategies. In: Gaur RK, Sharma P, editors. Approaches to plant stress and their management. New Delhi: Springer; 2014. p. 173–90.
Suri VK, Choudhary AK. Effect of VAM fungi and phosphorus application through STCR precision model on crop productivity, nutrient dynamics and soil fertility in soy- bean–wheat–soybean crop sequence in an acid Alfisol. Commun Soil Sci Plant Anal. 2013;44:2032–41.
Talnikar AS, Kadam GL, Karande DR, Jogdand PB. Integrated weed management in pigeon pea (*Cajanus cajan* (L.) Millsp.). Int J Agric Sci. 2008;4:363–70.
Taurian T, Castro S, Fabra A. Physiological response of two peanut rhizobia strains to acid pH. Symbiosis. 1998;24:327–36.
Taylor AG, Harman GE, Nielsen PA. Biological seed treatments using *Trichoderma harzianum* for horticultural crops. Hort Technol. 1994;4:105–9.
Tayyab AM, Qasim M, Azeem M, Ahmed N. Salt stress responses of pigeon pea (*Cajanus cajan*) on growth, yield and some biochemical attributes. Pak J Bot. 2016;48:1353–60.
Tejera NA, Soussi M, Lluch C. Physiological and nutritional indicators of tolerance to salinity in chickpea plants growing under symbiotic conditions. Environ Exp Bot. 2006;58:17–24.
Tucker CM. Pigeon pea anthracnose. Aust J Agric Res. 1927;34:589–96.
Umesha C, Sridhara CJ, Kumarnaik AH, Shivarajkumar HS. Ways to bridge yield gaps and production problems in pigeon pea cropping systems. J Pharmacogn Phytochem. 2017;6:2651–7.
Upadhyay RS, Rai B. Studies on antagonism between *Fusarium udum* Bulter and root region microflora of pigeon pea. Plant Soil. 1987;101:79–93.
Vadez V, Krishnamurthy L, Serraj R, Gaur PM, Upadhyaya HD, Hoisington DA, et al. Large variation in salinity tolerance in chickpea is explained by differences in sensitivity at the reproductive stage. Field Crops Res. 2007;104:123–9.
Verma CB, Lallu B, Yadav RS. Effect of boron and zinc application on growth and yield of pigeon pea. Indian J Pulses Res. 2004;17:149–51.
Vishwa D, Chaudhary RG, Mishra S, Khan AA. Occurrence of pigeon pea wilt caused by *Neocosmospora vasinfecta*. Indian J Pulses Res. 2005;18:254–5.
Zahir ZA, Arshad M, Frankenberger WT. Plant growth promoting rhizobacteria: applications and perspectives in agriculture. Adv Agron. 2003;81:97–168.

Chapter 10
The Vital Foliar Diseases of *Cicer arietinum* L. (Chickpea): Science, Epidemiology, and Management

Udaya Kumar Vandana, Naseema Hayat Barlaskar, Rijusmita Kalita, Islamul Hoque Laskar, and P. B. Mazumder

10.1 Introduction

Chickpea production in the world has increased over the past two decades, ranking third after dry bean (*Phaseolus vulgaris* L.) and field pea (*Pisum sativum* L.) (Hirdyani 2014). It dominates other legumes in the international market, and its trade traffic is more than 8 billion dollars annually (Stagnari et al. 2017). This crop contributes to agricultural sustainability through N_2 fixation and allows agricultural production by diversification. India is also the largest chickpea-producing country with 9.33 million tonnes production in 9.48 million ha of cultivated areas (Pande et al. 2005). The productivity in India is lesser in comparison to other chickpea-producing countries because of the biotic and abiotic stresses and also due to fungal foliar diseases. Chickpea is grown commercially in soils having residual moisture and with or without minimum irrigation in RRFL (rainfed rice fallow lands) (Pande et al. 2012). The optimal conditions needed for growth and development of chickpea include temperature around 18–26 °C during the night and 21–29 °C during the day and a total of 560–660 mm of annual rainfall. Chickpea is broadly classified into two types: desi type and kabuli type. Desi-type chickpea has seeds that are small and have sharp angular edges, and the color of the seed varies from black to almost cream color or yellow. The desi-type flowers are pink in color and produce about 80–90% of the chickpea throughout the world. *Dal* (the splits) and *besan* (flour) are made up of desi type (Purushothaman et al. 2014; Toker et al. 2007). The kabuli type has large, rounded seeds that are head-shaped having cream beige seed color and white seed coats (Pande et al. 2012). Production of chickpea is constrained by foliar diseases as well as insect pests. In general, fungal foliar diseases like Ascochyta blight, Botrytis gray mold, etc. are spread in northern, northern-western, and eastern India (Bretag et al. 2008).

U. K. Vandana · N. H. Barlaskar · R. Kalita · I. H. Laskar · P. B. Mazumder (✉)
Department of Biotechnology, Assam University, Silchar, Assam, India

B. P. Singh et al. (eds.), *Management of Fungal Pathogens in Pulses*,
Fungal Biology, https://doi.org/10.1007/978-3-030-35947-8_10

10.2 History and Origin: Chickpea

Chickpea is a historical crop of the modern age; it was cultivated since 9500 years ago in the Fertile Crescent, through Turkey to Iran (Harlan 1971). Chickpea is cultivated in association with other crops like wheat, pea, barley, lentils, flax, and vetch as a part of agricultural evolution in the Fertile Crescent (Abbo et al. 2003a). The large area spreading over Israel to Western Iran, from southeast Turkey to Jordan and Iraq, ascertained a balanced collection of basic needs like carbohydrate, protein, oil, and fiber (Diamond 1997). Wild plants were cultivated primarily in this region and were observed archaeologically, and information from 7500 BC and recent years remain feasible (Fuller and Harvey 2006). Chickpea is used as a food in the eighth millennium BC (Tanno and Willcox 2006). Even though, archeological records in chickpea are scarce because the seed is almost crushed down in the carbonization of seed Neolithic chickpea supported the distribution which restricted during the Fertile Crescent, especially at Anatolia and the eastern Mediterranean (Van der Maesen 1972). Later, the Neolithic Period, chickpea expanded westward to modern Greece. During the Bronze Age, chickpea has been spread widely to the west of Crete, south of Upper Egypt, eastward through recent Iraq toward the Indian subcontinent, where the other was found in Harappan community in Pakistan and various sites in Maharashtra and Uttar Pradesh (Colledge et al. 2004). During the Iron Age, chickpea was spread in South and West Asia and in Ethiopia. The crop expanded with the group of originator crops from the Fertile Crescent toward West Central Asia and also Europe from 5500 BC (Moreno and Cubero 1978). In the sixteenth century AD, chickpea was produced by the Spanish region and Portugal; and in the eighteenth century, kabuli type spread in the Indian region from the Mediterranean region (van der Maessen 1972). Indian immigrants in the later nineteenth century imported the desi chickpea to Kenya (van der Maessen 1972). At present in the USA, Canada, and Australia, chickpea breeding programs have started. The related species of chickpea is *Cicer reticulatum*, which is the only related species in the gene pool and spread in southeast Turkey. Numerous additional *Cicer* species of almanac and perennial are hereditarily found in the genetic makeup as per AFLP (amplified fragment length polymorphism) analysis (Kumar et al. 2016). The actual difference among the wild relatives and the native chickpea is the loss due to vernalization which is a polygenetic attribute (Abbo et al. 2003a). The most widespread production of chickpea occurs in North America and the Middle East and un-moistured winter regions of India (Abbo et al. 2003b).

10.3 Center of the Diversity of Chickpea

The spread of old and wild type occurs in the main three areas from 8° N to 56° N latitude and 8° W to 85° E longitude especially Ethiopia, Crete, Western Mediterranean, Greece, the Caucasus Iran, Asia Minor, Central Asia, Himalayan region, and Afghanistan. Domestic chickpea is presently highly nurtured in Australia, southern South America, African Mediterranean regions, Ethiopia, the European Mediterranean region, southern Asia toward Iran to Myanmar, and the Middle East encompassing Turkey, Iraq, and Israel (Van der Maesen 1972). International Crops Research Institute for the Semi-Arid Tropics (ICRISAT), India, is the largest GenBank for chickpea, which consists of 17,250 accessions and 6390 of Indian diversity, followed by 4850 of Iran, 930 of Ethiopia, 700 of Afghanistan, 480 of Pakistan, 470 of Turkey, 390 of Mexico, 220 of Syria, 139 of Chile, 133 of Soviet Union, and many additional countries from Northern Africa, Southern Europe, East Africa, North America, and South America (Abbo et al. 2003a). International center for Agricultural Research in the Dry Areas (ICARDA), Aleppo, Syria, for chickpea of kabuli type, the genebank consists of 12,070 accessions from 1780 of Iran; 970 of Turkey; 410 of India; 340 of Chile; 300 of Uzbekistan; 280 of Spain; 270 of Tunisia; 230 of Morocco; 210 of Bulgaria; 170 of Portugal; 160 of Russian Federation; 160 of Mexico; 150 of Jordan; 120 of the USA; 110 of Bangladesh, Tajikistan, and Azerbaijan; and some further provinces lesser (100) like Italy, Ethiopia, Palestine, South America, Algeria, North Europe, tropical Africa, and Egypt (Diamond 1997).

10.4 Chickpea Production

Chickpea is also known by different local names: hamas (in the Arab world), zimbra (in Ethiopia), nohud or lablabi (in Turkey), chana (in India), and garbanzo (in Latin America). Chickpea crop production spread from 6.6 million tonnes in the year 1998–1999 to 9.5 million tonnes by 2000–2001 (Moreno and Cubero 1978).

10.5 Ecology of Chickpea

The chickpea evolution is different from the other wild type of the West Asian Neolithic crops, and it shows a part in regulating the crop habitat. The chickpea habitat can be characterized easily with the advanced high-resolution information of the climate and geographical information system (GIS) software freely present in the public databases (Hijmans et al. 2001). The areas like Egyptian Nile Valley, Iraq, Pakistan, and central Iran retain the lowermost annual report precipitation in cold winter and mix the midsummer heats with adequate winter temperatures (Rousta et al. 2018).

- *Temperature and altitude of chickpea*
  Altitude and rainfall variableness remain less in Europe than in West Asia and North. South Asia's yearly temperatures remain higher, earlier to the beginning of the monsoon. It was observed minor dissimilarity between the mean

temperature of the warmest quarter in between northern and southern halves of the subcontinental distribution of chickpea which ranges from 30.8 to 31.9 °C, and the mean winter temperatures vary from the North (16.8 °C) to the South (22.1 °C). In Central Asia chickpea is cultivated in areas with a high series of temperature variation and rainfall unevenness, which leads to a hasty change in the altitude region (Bhat et al. 2017).

- *Summer-dominant rainfall region environments of chickpea*
  Chickpea-growing regions like South Asia, Peru, and East Africa are summer-dominant rainfall environments (Ahmed et al. 2016). There is a strong decrease in rainfall in the Indian subcontinent from the southeast to northwest; Madhya Pradesh, a central state with higher rainfall and summer-dominant rainfall region in the subcontinent, produces 50% of the chickpea (Bhat et al. 2017). Chickpea growing region in Mexico is arid from 119 to 284 mm/year and is a summer-dominant rainfall region, where summer and winter rainfall proportion increases from 36% to 43% and 33% to 46%, respectively (Nicholson 2014).

10.6 Adaptation of Chickpea: Stresses, Cropping Systems, and Habitats

10.6.1 Stresses in Chickpea

Stresses in chickpea can be classified as biotic stress and abiotic stress based on coarse agro-climate divisions. In the Mediterranean rainfall region and summer-dominant rainfall region, drought is dangerous and is intensified by heat pressure (Saxena et al. 1996). In India, for most summer-dominant rainfall region, *Fusarium* wilt-root rot complex, Ascochyta blight, and Botrytis gray mold are the biotic stresses which contribute to disease distribution and are estimated to cause 10% of the annual yield loss (Singh 1990).

10.6.2 Cropping Methods of Chickpea

Seeding methods for chickpea vary in various environments. The highest range of seeding approaches was found to be in the Mediterranean region, because of the comparative strength of the biotic and abiotic stress (Rasool et al. 2015).

Maturation of Chickpea in Late Spring or Early Summer of the Autumn-Sown Rainy Season

It is a regular chickpea cultivation system for regions with relatively warm winter and less biotic stress or pressure because it works on intra-seasonal rainfall and decreases the disclosure to drought. In West Asia and North Africa (WANA), especially in warmer regions of Iran and in the Nile Valley, both countries use this system to grow chickpea with supplementary irrigation in the areas to decrease drought stress (Saccardo and Calcagno 1990). At present-day winter, sowing and drill irrigation have been used by approximately 90% of Israeli farmers. Australia is biotic stress-free until the mid-1990s, and later production of chickpea declined, but it is recovered by the release of resistant variety and by adopting good management practices (Hughes et al. 1987). In Mediterranean Australia, winter temperature is moderate, and autumn sowing of chickpea is exposed to suboptimal temperature on flowering and can delay pod set by 30 days more. Prompt flowering expands yield constancy and attains alteration to water deficiency but expands the threat of encompassing less temperature (Saccardo and Calcagno 1990).

Spring-Sown Chickpea in Post-Rainy Season Maturation in Summer
It is a regular chickpea cultivation system of Mediterranean climates in WANA and minimizes the risk factor of winter frosts, disease stresses, and the farmers to take a decision for planting based on soil moisture profile (Hamwieh and Imtiaz 2015). In Tunisia, winter stress is lowered geographically, as the crops are grown in low elevation of <600 m deep clay loams in areas of semiarid, avoids heavy rainfall of >1000 mm/year, along with areas and frost-prone areas. Chickpea is sown in the middle of May to escape high temperatures which will occur post-October in north-eastern Australia (Saccardo and Calcagno 1990). Cultivation can be done in dry-sown if sufficient soil moisture is present, or as farmers delay for opening rain, which leads postponement of sowing till August in a few of the regions (Hamwieh and Imtiaz 2015). The chickpea crop can tolerate heat stress at phases of maturation, normally in November (27–30 °C). Although rainwater tends to rise from October in many regions, chickpea crop cannot enter the similar terminal drought stress in South Asian environments (Kumar 2017).

10.6.3 Chickpea Habitat Range

Chickpea is grown in diverse habitat which consists of altered climate, cropping system, and stress. Chickpea is essentially separated into definite ecotypes, showing local selection pressures in the region of millennia. From the past 30 years, there has been an evaluation of germplasm ranging from characterization and resistance screening by many international centers and physiological studies based on accession number (Upadhyaya 2003). Chickpea physiological and habitat understanding is a must, and major stresses can be avoided by combinations of sowing strategies and appropriate phenology (Berger et al. 2006). Chickpea phenology is increased by drought stress as it decreases the thermal time for flowering, maturity, and pod fill; it also lowers the water potential, photosynthesis pod number, and yield (Berger et al. 2006). Chickpea is also having dehydration postponement and consistency of tolerance, like deep rooting, extraction of high soil water, and adjustment of osmosis (Summerfield et al. 1985). Chickpea is highly tolerable to heat stress than the other cold grain legumes like field pea, lentils, and faba bean, and it also absorbs less incident radiation approximately <50% photosynthetically available radiation (PAR) than other seasonal legumes. Kabuli-type characters were demonstrated in East Asia, Europe, and the Mediterranean, while the desi character was common in Africa and also in Southeast Asia (El-Amier et al. 2015). The vegetative phase, when extended under long-season conditions, increased biomass accumulation and reproduction and delayed flowering till the temperature becomes sufficiently warm to aid the pod set. The difference among provinces has been detected in the assessment of Ethiopian drought-tolerant germplasm (Berger et al. 2006).

10.7 Uses, Consumption, and Utilization

From the beginning of agricultural time, legume crops have several uses depending on the utilization of different plant parts. Dry or green seeds are applicable for animal feed, fodder, and organic manure. It is also used as a whole and mixed with other cereals (Kumar 2017). Legumes are eaten as a main course in the dish either singly or with meat, fish, and snacks, green or dried. One of the examples of legume is chickpea. Chickpea can be packaged, ice-covered, canned, and precooked. It is a source of oil which is used in baking protein-rich cake (Venn and Mann 2004). It contains protein and carbohydrates and has nutritive value. It can also fix nitrogen from the atmosphere which is secreted into the soil. The cultivation is decreasing in recent years due to the cause of their marginalization of a late entry in the market (Rimal et al. 2015). The legume crop is the essential food of the vegetarian dietary system, so it is directly linked with Indian civilization (Agbola et al. 2002). The pulses or legumes can be dried properly and conserved to consume throughout the year. Consumption per capita of pulses of 80 g/day is advised by the World Health Organization (WHO) and consumption of 47 g/day by the Indian Council of Medical Research (ICMR) (Misra et al. 2011). Consumption in India is less than 30–34 g/day/person because of the unavailability and price rise of pulses (Akinjayeju and Ajayi 2011).

10.8 Nutritional Value of Chickpea

Nutrition through food is necessary for human life. Nutrition provides energy, macronutrients, micronutrients, etc. for growth, tissue maintenance, regulation of metabolites, and physiological functions. Chickpea in many countries is a staple food and plays an important element in the diet of vegetarians around the world. Chickpea is a valuable source of minerals, vitamins, energy, fibers, and also health-beneficial phytochemicals (Brenes et al. 2008).

Nutritional Composition

The nutritional composition can vary due to the environment, climate, soil biology, soil nutrient, stress factors, and agronomic factors (McCleary 2003).

- *Energy*
  Energy is defined as gross energy (MJ/kg) or as a caloric value (kcal/100 g). Chickpea has an energy value of 14–18 MJ/kg or 334–437 kcal/100 g for desi types, and for kabuli type, it is 15–19 MJ/kg or 357–446 kcal/100 g. It showed that the kabuli type has higher energy than the desi type due to the presence of the seed coat component (Perttilä et al. 2005).
- *Protein and amino acid*
  The protein concentration of desi type ranges from 16.7% to 30.6% and for kabuli type 12.6% to 29.0%. Chickpea is used for the treatment of malnutrition and kwashiorkor in children because of its high protein content (Greenfield and Southgate 2003). The body is also provided with amino acids to synthesize new proteins for repairing and replacing damaged tissue and to synthesize enzymes, hormones, and growth factors. Chickpea has a high amount of sulfur amino acid than the lysine (Sotelo et al. 1987).
- *Lipid and fatty acid*
  Chickpea consists of 2.9–7.4% lipid content for desi and 3.4–8.8% content for kabuli (Jukanti et al. 2012). The total lipid content consists of 62–67% of polyunsaturated, 19–26% of monounsaturated, and 12–14% of saturated fatty acids. Essential fatty acids like linolenic and linoleic acid are supplied through the diet (Trumbo et al. 2002).
- *Carbohydrates*
  Carbohydrates are the most important component in chickpea, having 54–71% for desi type and 54–71% for kabuli type (Greenfield and Southgate 2003). The key types of carbohydrates present are oligosaccharides (like raffinose (2.2%), stachyose (6.5%), ciceritol (3.1%), and verbascose (0.4%)), polysaccharides (like starch (30–57%)), monosaccharides (like glucose (0.7%), ribose (0.1%), fructose (0.25%), and galactose (0.05%)), and disaccharides (like maltose (0.6%) and sucrose (1–2%)) (Joint FAO/WHO 1998).
- *Minerals*
  Chickpea plants absorb the minerals (like, B, Fe, Mn, Zn, Cu, Ni, Ca, Mg, K, P, S, Cl, and Mo) from the soil and transfer it to the seed performs a metabolic activity like photosynthesis, respiration, chlorophyll synthesis, and cell division (Sujak et al. 2006).
- *Vitamins*
  Chickpea comprises of a high source of water-soluble vitamins like the B-complex vitamins (B1, B2, B3, and pantothenic acid) and vitamin C and lipid-soluble vitamins like vitamin A (provitamin A carotenoids), vitamin E (tocopherols and tocotrienols), and vitamin K (Who and Consultation 2003).

10.9 Foliar Fungal Disease

Chickpea is the most essential cool season pulse crop grown in dry regions. The chickpea plant agonizes commencing fungal foliar diseases that distress the growth stage of chickpea. The pathogens that infect the plant include bacteria, nematodes, viruses, fungi, and mycoplasma, which lead to severe crop yield loss. Among this, fungi are the most threatening group that affect the roots, stems, flowers, leaves, and pods of chickpea (Nene et al. 2012).

*10.9.1 Ascochyta Blight (***Ascochyta rabiei** *(Pass.) Labr.)*

Distribution
Ascochyta blight (AB) is a viral disease found in West Asia, Southern Europe, and Northern Africa. In Pakistan, it occurs in February and March and disease will develop accordingly; and in Northern India, it happens when the crop canopy is very dense. In West Asia, Northern Africa, and Southern Europe, such situations usually occur from March to May. In winter chickpea is sown toward the Mediterranean region, and the blight symptoms are found when the climate is wet and warm in November and December. The disease has been found to develop among 35 countries along 6 continents and presently seen in Canada and Australia; it can expand swiftly to different areas of chickpea production (Nene et al. 2012).

Economic Importance
The fungal foliar disease causes crop yield loss and quality loss of up to 100% (Nene et al. 2012).

Epidemiology
Ascochyta blight occurs through seed transmission of *Ascochyta rabiei*. Airborne spores of *A. rabiei* are found to play a major vital role in epidemics of the disease (Kaiser et al. 2000). *A. rabiei* either lives on the seed or inside it or can be found in the plant debris of diseased left over in the fields as a mycelium and pycnidia or at its teleomorph stages and can serve as an agent of the disease (Santra et al. 2001). The secondary spread of this fungus occurs through conidia and ascospores. Development of teleomorph, the stage of the sexual reproductive, appears due to the mating of compatible new types in new areas spread through the air (Guarro et al. 1999). The teleomorph stage assists the pathogen in a longer duration of survival in its host, though it has never been seen in the newly infected host. In many regions, though, pseudothecia are found in infected plant wastes. Seed transmission in a field causes pathogen distribution randomly, giving the cause of many initial infections. Wet, cloudy, and cool weather is favorable for the disease development. In a cool climatic condition, the density of asci and ascospore production per pseudothecium are much higher than the warm condition (Daehler et al. 2004). Ascospores are also necessary for dispersal of the pathogen to long distances. The ascospore gets discharged to the air from pseudothecium during the wet condition. Production of ascospore on largely infected crop residues can reach up to 1.5×10^4 ascospores/mm^2 on the tissue surface (Manstretta and Rossi 2015). The productions of conidia per pycnidium are much more in cool regions compared to warmed counterparts. Strong wind and rain can scatter conidia grown on diseased plant parts, provided if conidia are present in water droplets or rain splash. Relative humidity compared to temperature plays a more vital role as a critical factor in the determination of the

development of pseudothecia and pycnidia on crop debris (Vidal et al. 2017). The disease best develops at low temperature, optimum being at 20 °C. The moist environment also acts as a vital factor to produce severe infection. Dry periods after immediate inoculation may sometime induce disease severity though dry period exceeding 12 hours after 6 hours of wet treatment may reduce the disease development. In tropical countries, *A. rabiei* by crop debris get influenced by the low rainfall and high temperature during the out of season summer months, which is detrimental for the survival of the pathogen *A. rabiei*. Impacts of light in in vitro conditions reportedly have insignificant influence on pseudothecial development and discharge of ascospores (Sehulster and Chinn 2003).

Symptoms

AB is typically seen during the flowering and podding stage as patches (Gurjar et al. 2012). The disease can be observed at an early stage of growth. When the pathogen is seed-borne, the germination time is favorable for the development of disease at the stem base with dark brown lesions (Lammerts van Bueren et al. 2004). The seedlings which are affected can be collapsed and die due to the formation of pycnidia. The disease spread from the seedling to the flowering and podding which results in patches of diseased plants. The disease appears in the form of spots of small water-soaked in the young leaves in the branches when the origin is airborne and conidia or ascospores (Nene et al. 2012). These spots enlarge and integrate which blights the leaves and the buds that lead to disease development under favorable conditions and also pycnidia presence on blighted leaves and buds. Because of susceptible cultivation, the necrosis spread through the buds, which kill the plant. In severe infection of the foliar disease, the entire plant gets dry and falls off. If the temperature is hot, the condition is unfavorable for a disease formation, and the infection remains in the leaves, stems, pods, and petioles as discrete lesions. The symptoms appear like round spots that have brown margins where pycnidia are presently showing a gray center that appears like concentric rings. Lesion size varies from 3 to 4 cm long on stems. If the disease arises during the pre-flowering stage when conditions are unfavorable, the crop grows with the symptoms that are visible on the older branches. Pods with fully developed lesions are round having 0.5 cm diameter along with pycnidia arranged in concentric rings. The pod becomes blighted and fails to grow any seed if infection occurs in the early developmental stages of the pod growth. Shriveled seed and infected seed have resulted from late infection. The seed shows symptoms of brown discoloration and visible pycnidia which can be seen by the naked eye (Pande et al. 2012).

Pathogenesis

Ascochyta rabiei germinates after 12–48 hours of inoculation. Through leaflets, the pathogen reaches to petiole and then attacks the stem. Following its germination, the pathogen forms its germ tube and appressorium-like structure, which is a specialized hyphal cell that occurs at the tip toward the germ tube required for penetrating the plant cell. The appressorium is kept apart from the germ tube through a septum and surrounded by mucilaginous exudates. The fungus at first penetrates its hyphae through the cuticle and traversing the subcuticular region reaches the forefront of epidermal cells. Penetration in the epidermal cells occurs through the wall, keeping the protoplasmic structures intact, reaches to the intercellular space, and resides and grows between epidermal and palisade parenchymal cells (Pande et al. 2005). The diameter of hyphal cells varies in and out of the cell as 3.5 μm and 2 μm, respectively. Meanwhile, dark aggregates of mycelia start to grow at the subepidermal portion. Subsequently, the structure of epidermal, palisade, and spongy parenchyma starts to deteriorate and eventually gets disorganized. Infection near the stoma occurs through penetration of hypha through the juncture of guard cells and subsidiary cells regardless of whether the stoma is open or close. After the disorganization of leaf cells, pycnidium emerges from the damaged tissues. From pycnidium, conidiophores arise and subsequently conidium gets dispersed into the surrounding environment and through which new chickpea crops get infected (Galloway and MacLeod 2003). The pycnidia originate after the fifth day of inoculation. By the seventh day, non-lignified cells almost get deteriorated particularly through necrosis, but lignified cells like xylem and tracheary elements remain mostly unharmed. The pathogen while spreading from leaflet to stem through petiole infects the phloem vessels with less or no harm to xylem vessels, and consequently in some instances, the leaf breaks off from petiole. However, the fungal hyphae colonize both the xylem vessels and phloem vessels in the stem, and the walls of xylem and phloem vessels remain intact, while extensive damage happens to parenchymatous tissues (Smith et al. 2017). Although pathogen infects stems directly through its cuticle evading the usual route from the leaf, during pycnidia formation, parenchymatous cortical degradation and tissues of the pith degradation suggest that involvement of toxins and enzymes for cell wall digestion is inevitable (van den Brink and de Vries 2011). Reportedly, in the process of the pathogenesis of *A. rabiei*, solanapyrone A, solanapyrone B, and solanapyrone C are required. Though under in vivo condition only solanapyrone C has been found and nonappearance of other toxins in experimentation probably due to their low concentration. The application of solanapyrone in combination or independently results in prominent symptoms followed by an epidermal, palisade, and spongy parenchymal tissue contraction due to the effect of toxins in the protoplasm. Solanapyrone A is said to be the most toxic, resulting in shriveling, loss of turgor, broken stem, and chlorotic leaves (Kim et al. 2015). Phytoalexins like pterocarpans get degraded by *A. rabiei* through its conversion to 2-OH isoflavones and 1a-OH pterocarpans due to the activity of reductase and hydroxylase enzymes. The two kinds of enzymes particularly act upon two isomeric forms of phytoalexins, namely, maackiain and medicarpin. Apart from these enzymes, cutinase and polygalacturonase are also found to act upon the host system (Uchida et al. 2017).

*10.9.2 Botrytis Gray Mold (*Botrytis cinerea *Pers. ex Fr.)*

Distribution

Botrytis gray mold (BGM) is a foliar disease found in Bangladesh, Nepal, India, Pakistan, Argentina, and Australia. BGM has also been observed in Canada, Chile, Mexico, Hungary, Spain, Turkey, Vietnam, and the USA (Jain 2011).

Economic Importance

BGM fungal foliar disease causes yield losses of about 10% (Tivoli et al. 2006).

Epidemiology

Botrytis gray mold is the most detrimental crop disease after Ascochyta blight (Shafique et al. 2014). The pathogen of this foliar disease has a very high host range and can live on other crops as well as weeds, and hence the disease is widespread. Damages mostly occur during higher temperatures and humidity. The temperature required is greater than that of the optimum temperature needed for Ascochyta blight development. BGM originates from seed, and the fungus has a large range of hosts. The disease is generally observed during floral growth when the canopy of the crop is fully matured. Excessive vegetation, too much irrigation, rain, and close spacing are causes that favor disease growth and development. Temperature ranges between 20 and 25 °C and high humidity during podding and flowering period also favor disease growth. The disease may also occur subsequently after the appearance of Ascochyta blight (Malhi et al. 1994). *Botrytis cinerea* can inhabit on chickpea seed without showing any symptoms for more than 5 years. The period of survival gets largely affected by the storage temperature, particularly between 5 and 10 °C being optimum for survival for up to 5 years. The temperature at 20 °C has been observed to have reduced growth of the pathogen from 95% to 2% at the duration of 12 months. Heating the infected seed at the moist condition at a temperature of 50 °C resulted in a significant reduction of the infection (Williamson et al. 2007). Studies showed that chickpea leaves infected with the fungus get decomposed within a couple of days to months, but the deterioration of stems through infection requires longer duration. In India, the pathogen is observed to survive for approximately 8 months in leftover infected crops on the soil and is the principal source of the initial inoculum. Asexual sporulation of the pathogen occurs on the stubble during higher temperatures and high humidity. Spores get blown to the air from the debris of the infected crop and spread to other places. The pathogen inhabits the soil in the form of mycelia and sclerotia (Bhaskar et al. 2009). In crop stubbles, sclerotia occur in many host species, as the disease has long-term survival on the host. However, in Australia, sclerotium does not show long-term survival. In Europe, apothecia originate from fertilized sclerotia (Cannon and Kirk 2007). Chlamydospore occurs during extreme conditions like drought, nutrient deficiency, bacterial attack, and change of pH. Mycelium can be produced through the germination of chlamydospores, which serves as secondary inoculum (Stevens 2002).

Symptoms

The absence of pod setting is the primary symptom of the disease where leaves and stems do not show symptoms. The disease shows symptoms under highly favorable conditions and forms patches in the plant which often dies. The symptoms are visible on stems, pods, leaves, and flowers as a dark brown or gray lesions layered with sporophores under high humidity. 10 mm- to 30-mm-long lesions are present on the stems which grid the stem fully. The branches break at the place of the gray mold where it has caused rotting. The leaves and flowers which are affected become a rotting mass. Lesions become water-soaked and shaped irregular on the pod.

The pod consists of small and shriveled seeds or a lack of seed in the infected plants. In the infected seeds, grayish-white mycelium is observed (Narayanasamy 2011).

Pathogenesis

The spore of *Botrytis cinerea* germinates after 6–8 hours of inoculation. The fungus *B. cinerea* being a necrotrophic organism grows saprophytically on the leaf. The germ tube develops and forms a mycelial connection on the leaf. The tip of the germ tube forms appressorium, necessary for penetrating the plant cells. The pathogen penetrates the host system through the cuticle of leaf and resides and formation of mycelium at subcuticular or subepidermal layer. The penetration through stomata has been observed in the spore of *Botrytis cinerea* which germinate after 6–8 hours of inoculation. The germ tube develops and forms a mycelial connection on the leaf. After establishing itself at subcuticular or subepidermal position, the hyphae grow and reach to mesophyll cells. The hyphae thicken and start branching at the mesophyll layer, consequently damaging mesophyll and epidermal cells. The degradation of the two layers requires cell wall enzymes such as pectinases, cutinases, cellulases, and polygalacturonases. As the pathogen cannot degrade lignin, it does not affect the lignified cells like xylem and tracheary elements. The degradations of mesophyll cells occur after 72–96 hours of inoculation. The total necrosis of the leaf takes place after 120 hours of inoculation, and characteristic yellowing of the leaf is observed (Arranz et al. 2000). The reactive oxygen species (ROS) can be generated by *B. cinerea* during its metabolic processes or with the help of NADPH oxidases (NOX). The NOX is a protein of muti-subunit and can reduce superoxide anion from oxygen. The BcNoxA and BcNoxB are catalytic subunits of NOX; BcNoxA helps pathogens to colonize on host tissues, whereas BcNoxB is necessary for primary infection. Apart from these two subunits, another regulatory subunit BcNoxR is responsible for the growth, sporulation, and increased virulence of the pathogen (Hua et al. 2018). Cell wall enzymes are necessary for degrading the structural polysaccharides of the host cells. Cutinases are responsible for degrading cuticles and cellulases for cellulose. Endo-β-1,4-xylanases and pectin methylesterases found in the cell wall are necessary for degrading xylan and dimethyl esterification of cell wall components like polygalacturonase, respectively, and therefore endorse the pathogen into its entry to host environment. Two endo-polygalacturonases, BcPG1 and BcPG2, are required for virulence of the pathogen. Both BcPG1 and BcPG2 are necessary for primary infection, while BcPG2 is also involved in lesion expansion (Ten Have et al. 2010).

10.10 Management

10.10.1 Host-Plant Resistance

Host-plant resistance can be termed as the adaptation taken from different herbivores or pathogens for improvement in reproduction and sensitivity. Plants are sensitive; they produce several allelochemicals (secondary metabolites) which have been used by the plant to inhibit the growth, behavior, and survival of different pathogens (Pande et al. 2006). Pathogen inhibition can be also triggered by hypersensitivity (HR), reinforcement of cell wall by deposition of lignin, callose glycoprotein which is rich in hydroxyproline, polyphenols or cinnamic acid, etc. against leaf cuticle thickening parasite by epithelium thickening, which provides a mechanical barrier. In the case of a disease like Ascochyta blight, resistance is also induced by increasing the respiration rate and carbohydrate content of second days after inoculation (DAI). It has resulted in a hypersensitivity response. Second DAI gives resistance to ILC 32792 genotype by hypersensitivity response. Rather than hypersensitivity response, metabolic compounds like phytoalexin are involved in the exertion of defense mechanisms toward photogenic fungi. It had been found that when the crude culture filtrate (CCF) of the strain *A. rabiei* was applied, accumulation of medicarpin (phytoalexin) is increased in the culture. Accumulation of phenolic compounds like formononetin and biochanin A also helps in inducing plant defense. Studies show that defense-related enzyme like hydrolytic enzymes and phenylpropanoid pathway's enzymes also has their role in plant defense. Accumulation of β-1,3-glucanase and peroxidase in the cell wall causes the hydrolyzing of the cell of fungi. Ascochyta blight disease can be controlled by inducing HPR (host-plant resistance) (Waliyar et al. 2016). In the case of Ascochyta blight, there are several screening methods used in field and greenhouse conditions. Screening in chickpea germplasm by HPR shows a high level of resistance against BGM, by using this HPR, advanced chickpea breeding lines Australia evaluates BGM resistance germ lines. These lines equally give resistance against Ascochyta blight (AB) (Kumar et al. 2018).

10.10.2 Seed Treatment

In countries like Australia, Canada, Iran, the USA, etc., Ascochyta blight in chickpea had been reported due to infected seed which results in the low seed weight and discoloration. In the case of chickpea, blight-free seed productions are widely used in disease management (Sharma and Ghosh 2016). The selection of larger-sized seeds against smaller ones reduces the chances of blight disease as small-size chickpea seeds have a higher level of Ascochyta infections. Seed immersion in the hot water and chemicals like $CuSO_4$ solution, thiram, malachite green, etc. are used to treat chickpea. Again fungicide dressing in the seeds of chickpea improves the resistance as it halts the spore germinations and mycelial growth on the surface of the seed (Singh and Reddy 1996). But due to several factors like soil characteristics, weather condition, and plant growth inhibition, it is found that blight disease is not prevented against the phytotoxicity of fungicides which give adverse effect on seed germination. It has been reported that treating chickpea using thiram, tridemorph, imazalil, etc. causes the loss of vigor and hence is not practiced widely (Mohammed et al. 2017).

10.10.3 Culture Control Method

The main concept of disease management is to produce pathogen-free seed. Different practices like erect cultivars, manipulating in showing dates, etc. help in reducing different foliar diseases. Late sowing lowers the vegetative growth and thus reduces the disease incidence. To allow more aeration, wider row spacing is practiced in the crop field, and it reduces leaf wetness, relative humidity, etc. Thus it helps in the reduction of disease occurrence in plants. Another practice in the plants with compact and erect growth also helps in reducing diseases than that incuse of bushy spreading. Bushy spreading happened because of low aeration. By practicing all the above, we can reduce the disease incidence in chickpea (Heydari and Pessarakli 2010).

10.10.4 Cut-Twig Method

In the cut-twig method, test genotypes are grown in a plastic bag (45/30/5 cm) which is filled with vermiculites (4:1) and sterilized sand and placed in a glasshouse at 25 ± 2 °C with susceptible check H208/JG 62 used for artificial inoculums. 10–15-cm-long tender shoot of chickpea plant was cut with a sharp edge blade in the evening. It is transferred to the test tube by wrapping the course portion with a cotton plug containing fresh tap water. It inoculates in a test tube by the susceptible check (G543 or H208 OR L3.0). The symptoms start to appear 24 hours, and after 6 days, 100% mortality of susceptible lines can be seen (Udall and Wendel 2006).

10.10.5 Resistance Sources and Studies on Disease Management

In reducing the control of Ascochyta blight, foliar spray of chlorothalonil and benomyl was used for increasing seed height and yield (Bretag et al. 2008). In Australia, they used thiabendazole and thiram for treating the chickpea seed which increases the yield by up to 20%. Complete resistance was seen in using inoculation of pregerminated seed and in the seed coat (1995). Benomyl or sulfur is used for spraying the foliages (Hagedorn 1996).

In Australia, the host plants which are resistant used in the industries are the best for various conditions or option for controlling these diseases. Some of them use the pathogen-free seed to break off at least 3 years between chickpea crops in the same field. They keep it at least 500 m away from last year's crop in delaying sowing to applying fungicide sprayed many times. Crop management practices where emphasized to decrease or to reduce the damage occurred due to diseases. Pathogenicity is the step of pyramiding resistance genes into genetic makeup. The key component of disease management is host resistance. Fungicide dressings help to prevent the spore germination and to eradicate the fungus from the seed coat. Another method used is the crop rotation which helps in controlling the diseases (Salam et al. 2011).

10.10.6 Breeding for Disease Resistance

In single plant progenies and advanced breeding lines, they use field screening techniques and growth room for segregation. The deoxyribonucleic acid marker will encourage using an exotic source of disease resistance. ICRISAT has seen the growth of AB resistance lines in desi-type chickpea. From the diverse source, multiple crosses are produced to accumulate resistance gene (Serraj et al. 2003). Conventional breeding method NIFA-88 has been developed with the application of propineb, zineb, ferbam, etc. This method helps to reduce the secondary spread of AB in crops (Sarmah et al. 2012).

10.10.7 Biological Control

Studies show that the strains like *Trichoderma harzianum* Rifai and *Trichoderma viride* give antagonistic effect on the *B. cinerea*. The growth of *B. cinerea* on the hyphal tips is inhabited by *T. viride* species. Spraying of *T. viride* on the seeds helps in the germination of the seeds. The T15 strain of *Trichoderma* species is used as an effective biocontrol agent. *T. viride* and vinclozolin are found to be more effective with the application of fungicides. To produce artificial resistance, it is treated with *T. viride* and *Gliocladium roseum* (Monte 2001). This application is equivalent to that of seed treated with thiram. Compounds like essential oil production in the plants also reduce the infection of *B. cinerea* from 90% to 80%. These essential oils include cinnamon oil, clove oil, etc. The essential oil effect is studied by an automatic microtiter plate. Bacterial species like *Thymus zygis* and *Cymbopogon martini* help in the production of essential oil which is antagonistic against *B. cinerea* (Wilson et al. 1997). Different techniques are involved in the study of growth inhibition of fungi, and this includes the production of glyoxalate which helps to combat different diseases. The biological control of foliar disease also helps in disease management without applying chemicals to the crop field (Shamsi and Khatun 2016).

10.10.8 Resistance Sources and Disease Management

In reducing the control of Ascochyta blight, foliar spray of chlorothalonil and benomyl was used for increasing seed height and yield (Bretag et al. 2008). In Australia, they use thiabendazole and thiram for treating the chickpea seed which increases the yield by up to 20%. Complete resistance was seen in using inoculation of pre-germinated seed and in the seed coat. Benomyl or sulfur is used for spraying the foliages. In Australia, the industries use host-plant resistance as the best long-term administration for diseases. Some of them use the pathogen-free seed to break off a minimum of 3 years before sowing chickpea crops in the same field. They keep a distance of 500 m from last year's crop in delaying the sowing to applying fungicide spray for several times (Pande et al. 2005). Crop management practices were emphasized to minimize the damage caused by these diseases. Pathogenicity is also the method of pyramiding resistance genes into genetic materials. The key component of disease management is host resistance. Fungicide dressings help to prevent the spore germination and to remove the fungal infections from the seed coat. Another method used is the crop rotation which helps in controlling the diseases (Johansen et al. 2008).

10.10.9 The Genetic Basis of Host-Pathogen Interaction

In the case of BGM, the gene control resistance was reported in 1985. In this, parents F1 and F2 and their backcross generation BC1 and BC2 screening for resistance against BGM under epiphytotic condition. A single dominant gene Bor1 gives resistance to ICC 1069. The cross of ICC 1069 with BGM 413 and BGM 256 gives the ratio like 13 resistances is to 1 susceptible plant. It shows that the two epistatic interaction genes control resistance. Different studies on resistant varieties like ICC 1069, P 349, NEC 2451 and 2 susceptible genotypes JG 62 and T3 in India and Australia produced BGM resistance cross. The resistance in the entire three parents is controlled only by one single dominant gene. The F2 produces 15 resistances in 1 susceptible plant (Leroux et al. 2002).

10.10.10 Gene Plant Technology

Gene technology nowadays is used for crop/plant improvement. In the case of chickpea, gene plant technology is used to treat diseases infected by both AB and BGM. Production of antifungal metabolites by expressing different genes is one such kind of gene plant technology. Different antifungal proteins and hydrolytic enzymes like chitinase are also accumulated by gene plant technology which degrades the cell wall of fungi. In the case of kiwi fruit, the production of β-1,3-glucanase reduced symptoms of *B. cinerea* infection. In the case of alfalfa ferritin, an iron-binding protein is also produced which gives protection against oxidative damage of necrotic pathogen. The transgenic plant which consists of polygalacturonase-inhibiting protein (PGIP) gives resistance against *B. cinerea*. The PGIP works against the PG that is secreted by the pathogen against the plant cell wall. This PGIP is isolated from raspberry and kiwi fruit which is introduced in different plants by gene plant technology. QTL mapping is used to study Ascochyta blight disease in pea plants (Sagi et al. 2017).

10.10.11 Integrated Disease Management (IDM)

Integrated disease management is the technique that manages the disease and mitigates yield at the same time. It involves the cultivation of pathogen-tolerant genotype, application of diammonium phosphate in soil and of Carbendezim or Thiram in seeds, and wider row spacing (0.6 m) against foliar diseases like Ascochyta blight and BGM. It is reported that ICCL873 22 genotypes were controlled by chemicals of BGM, wider row spacing is used, and *T. viride* is sprayed on the genotype (Pande et al. 2006). The Nepal Agricultural Research Council (NARC) and Natural Resources Institute (NRI), UK, reported the increase of health by 400% after the IDM program (Pande et al. 2006).

10.10.12 Field and Control Environment Screening for Disease Resistance

Different techniques for screening are developed at different research centers for chickpea, and it gives artificial resistance against foliar diseases like Ascochyta blight. The field screening and control environment screening are two major screening methods standardized by the ICRISAT (International Crops Research Institute for the Semi-Arid Tropics) and ICAR (Bidinger et al. 2009) against AB. This involves the planting of test material in a 40 cm row space. It also involves independent cultivation that serves as the indicator or spirit line. In a cloudy day, the spores are incubated in the plants at flowering time, and infected debris are spread between rows. Again these inoculates are integrated during the dry weather for approximately 15 days. In these plants, no visible lesions are found. Again, in the environmental screening, air temperature is maintained at 20 ± 1 °C, 12 hours of photoperiod, etc. (Landa et al. 2001).

10.11 Conclusion

Chickpea is a quantitative source of carbohydrates, proteins, minerals, vitamins, and fibers. Chickpea also fixes atmospheric nitrogen and reduces the need for nitrogen fertilizers. The crops are affected by serious foliar diseases, which affect the development stages. Botrytis gray mold and Ascochyta blight are among the most prominent diseases of chickpea. New and suitable understanding of the science, ecology, distribution, symptoms, epidemiology, pathogenesis, economic importance, and integrated management or control measures of the major foliar fungal diseases of chickpea is studied or focused on this chapter. The foliar disease has restricted chickpea production in many countries; therefore integrated management or control strategies are needed to be adopted to prevent loss of crop and pulses. Investigation of the pathogen's genetic basis of host-pathogen interaction and identification of the host-plant resistance will help in improving or breeding a resistant variety of chickpea and will be useful to farmers and researchers. Damage caused by fungal foliar diseases can be reduced by using moderate integrated resistant cultivars with the strategies of agronomic management practices. The management practice will result in a better resistance for the host plant and will lead to greater opportunities for sustainable agriculture and maximum productivity. Agronomic options are added to management to decrease the damage which is caused by the pathogen.

References

Abbo S, Berger J, Turner NC. Evolution of cultivated chickpea: four bottlenecks limit diversity and constrain adaptation. Funct Plant Biol. 2003a;30(10):1081–7.

Abbo S, Shtienberg D, Lichtenzveig J, Lev-Yadun S, Gopher A. The chickpea, summer cropping, and a new model for pulse domestication in the ancient near east. Q Rev Biol. 2003b;78(4):435–48.

Agbola FW, Kelley TG, Bent MJ, Rao PP. Eliciting and valuing market preferences with traditional food crops: the case of chickpea in India. Int Food Agribus Manag Rev. 2002;5(1):7–21.

Ahmed AM, Tana T, Singh P, Molla A. Modeling climate change impact on chickpea production and adaptation options in the semi-arid North-Eastern Ethiopia. J Agric Environ Int Dev. 2016;110(2):377–95.

Akinjayeju O, Ajayi OF. Effects of dehulling on functional and sensory properties of flours from black beans (*Phaseolus vulgaris*). Food Nutr Sci. 2011;2(04):344.

Arranz M, Eslava A, Benito E. Pathogenicity factors in *Botrytis cinerea*. Rev Iberoam Micol. 2000;17(1):S43–6.

Berger J, Ali M, Basu P, Chaudhary B, Chaturvedi S, Deshmukh P, Dharmaraj P, Dwivedi S, Gangadhar G, Gaur P. Genotype by environment studies demonstrate the critical role of phenology in adaptation of chickpea (*Cicer arietinum* L.) to high and low yielding environments of India. Field Crop Res. 2006;98(2–3):230–44.

Bhaskar PB, Venkateshwaran M, Wu L, Ané J-M, Jiang J. Agrobacterium-mediated transient gene expression and silencing: a rapid tool for functional gene assay in potato. PLoS One. 2009;4(6):e5812.

Bhat MA, Romshoo SA, Beig G. Aerosol black carbon at an urban site-Srinagar, north-western Himalaya, India: seasonality, sources, meteorology and radiative forcing. Atmos Environ. 2017;165:336–48.

Bidinger F, Yadav O, Rattunde EW. Genetic improvement of pearl millet for the arid zone of northwestern India: lessons from two decades of collaborative ICRISAT-ICAR research. Exp Agric. 2009;45(1):107–15.

Brenes A, Viveros A, Centeno C, Arija I, Marzo F. Nutritional value of raw and extruded chickpeas (*Cicer arietinum* L.) for growing chickens. Span J Agric Res. 2008;6(4):537–45.

Bretag T, MacLeod W, Kimber R, Moore K, Knights E, Davidson J. Management of Ascochyta blight in chickpeas in Australia. Australas Plant Pathol. 2008;37(5):486–97.

Cannon PF, Kirk PM. Fungal families of the world. Wallingford/Cambridge, MA: CABI; 2007.

Colledge S, Conolly J, Shennan S, Bellwood P, Bouby L, Hansen J, Harris D, Kotsakis K, zdoan M, Peltenburg E. Archaeobotanical evidence for the spread of farming in the eastern Mediterranean. Curr Anthropol. 2004;45(S4):S35–58.

Daehler CC, Denslow JS, Ansari S, KUO HC. A risk-assessment system for screening out invasive pest plants from Hawaii and other Pacific islands. Conserv Biol. 2004;18(2):360–8.

Diamond J. Location, location, location: the first farmers. Science. 1997;278(5341):1243–4.

El-Amier YA, El-Halawany E, Haroun SA, Mohamud SG. Vegetation analysis and soil characteristics on two species of genus *Achillea* growing in Egyptian Desert. Open J Ecol. 2015;5(09):420–33.

Fuller DQ, Harvey EL. The archaeobotany of Indian pulses: identification, processing and evidence for cultivation. Environ Archaeol. 2006;11(2):219–46.

Galloway J, MacLeod W. *Didymella rabiei*, the teleomorph of *Ascochyta rabiei*, found on chickpea stubble in Western Australia. Australas Plant Pathol. 2003;32(1):127–8.

Greenfield H, Southgate DA. Food composition data: production, management, and use. Rome: Food & Agriculture Organization; 2003.

Guarro J, Gené J, Stchigel AM. Developments in fungal taxonomy. Clin Microbiol Rev. 1999;12(3):454–500.

Gurjar MS, Ali S, Akhtar M, Singh KS. Efficacy of plant extracts in plant disease management. Agric Sci. 2012;3(3):425.

Hagedorn D. Pea enation mosaic enamovirus: ecology and control. In: The plant viruses. Boston: Springer; 1996. p. 345–56.

Hamwieh A, Imtiaz M. Identifying water-responsive and drought-tolerant chickpea genotypes. Crop Pasture Sci. 2015;66(10):1003–11.

Harlan JR. Agricultural origins: centres and non-centres. Science. 1971;174(4008):468–74.

Heydari A, Pessarakli M. A review on biological control of fungal plant pathogens using microbial antagonists. J Biol Sci. 2010;10(4):273–90.

Hijmans RJ, Guarino L, Cruz M, Rojas E. Computer tools for spatial analysis of plant genetic resources data: 1. DIVA-GIS. Plant Genet Resour Newsl. 2001;127:15–9.

Hirdyani H. Nutritional composition of chickpea (*Cicer arietinum* L) and value added products – a review. Indian J Community Med. 2014;26(Supp 2):102–6.

Hua L, Yong C, Zhanquan Z, Boqiang L, Guozheng Q, Shiping T. Pathogenic mechanisms and control strategies of *Botrytis cinerea* causing post-harvest decay in fruits and vegetables. Food Qual Saf. 2018;2(3):111–9.

Hughes G, Keatinge J, Cooper P, Dee N. Solar radiation interception and utilization by chickpea (*Cicer arietinum* L.) crops in Northern Syria. J Agric Sci. 1987;108(2):419–24.

Jain R. Pulse expert: an expert system for the diagnosis and control of diseases in pulse crops. Expert Syst Appl. 2011;38(9):11463–71.

Johansen C, Bakr M, Islam MS, Mondal N, Afzal A, MacLeod W, Pande S, Siddique KH. Integrated crop management of chickpea in environments of Bangladesh prone to Botrytis grey mould. Field Crop Res. 2008;108(3):238–49.

Joint FAO/WHO. Carbohydrates in human nutrition: report of a joint FAO. Rome: WHO, FAO; 1998.

Jukanti AK, Gaur PM, Gowda C, Chibbar RN. Nutritional quality and health benefits of chickpea (*Cicer arietinum* L.): a review. Br J Nutr. 2012;108(S1):S11–26.

Kaiser W, Ramsey M, Makkouk K, Bretag T, Açikgöz N, Kumar J, Nutter F. Foliar diseases of cool season food legumes and their control. In: Linking research and marketing opportunities for pulses in the 21st century. Dordrecht: Springer; 2000. p. 437–55.

Kim W, Park C-M, Park J-J, Akamatsu HO, Peever TL, Xian M, Gang DR, Vandemark G, Chen W. Functional analyses of the Diels-Alderase gene sol5 of *Ascochyta rabiei* and *Alternaria solani* indicate that the Solana pyrone phytotoxins are not required for pathogenicity. Mol Plant-Microbe Interact. 2015;28(4):482–96.

Kumar P. Food and nutrition security in India: the way forward. Agric Econ Res Rev. 2017;30(1):1–21.

Kumar K, Sardana S, Singh M, Gautam N. Management of germplasm collections in chickpea. Int J Agric Environ Biotechnol. 2016;1(3):565–76.

Kumar M, Yusuf MA, Nigam M. An update on genetic modification of chickpea for increased yield and stress tolerance. Mol Biotechnol. 2018;60(8):651–63.

Lammerts van Bueren E, Ranganathan R, Sorensen N (Eds). Proceedings of the 1st world conference on organic seed, FAO, Rome, 5–7 July 2004, pp.1–5.

Landa BB, Navas-Cortés JA, Hervás A, Jiménez-Díaz RM. Influence of temperature and inoculum density of *Fusarium oxysporum* f. sp. *ciceris* on suppression of Fusarium wilt of chickpea by rhizosphere bacteria. Phytopathology. 2001;91(8):807–16.

Leroux P, Fritz R, Debieu D, Albertini C, Lanen C, Bach J, Gredt M, Chapeland F. Mechanisms of resistance to fungicides in field strains of *Botrytis cinerea*. Pest Manag Sci. 2002;58(9):876–88.

Malhi S, Nyborg M, Beauchamp E. Large granules, nests or bands: methods of increasing efficiency of fall-applied urea for small cereal grains in North America. Fertil Res. 1994;38(1):61–87.

Manstretta V, Rossi V. Effects of weather variables on ascospore discharge from *Fusarium graminearum perithecia*. PLoS One. 2015;10(9):e0138860.

McCleary BV. Dietary fibre analysis. Proc Nutr Soc. 2003;62(1):3–9.

Misra A, Singhal N, Sivakumar B, Bhagat N, Jaiswal A, Khurana L. Nutrition transition in India: secular trends in dietary intake and their relationship to diet-related non-communicable diseases. J Diabetes. 2011;3(4):278–92.

Mohammed A, Tana T, Singh P, Molla A, Seid A. Identifying best crop management practices for chickpea (*Cicer arietinum L.*) in North-Eastern Ethiopia under climate change condition. Agric Water Manag. 2017;194:68–77.

Monte E. Understanding *Trichoderma*: between biotechnology and microbial ecology. Int Microbiol. 2001;4(1):1–4.

Moreno M-T, Cubero J. Variation in *Cicer arietinum* L. Euphytica. 1978;27(2):465–85.

Narayanasamy P. Microbial plant pathogens-detection and disease diagnosis, viral and viroid pathogens. Dordrecht: Springer Netherlands; 2011.

Nene Y, Reddy M, Haware M, Ghanekar A, Amin K, Pande S, Sharma M. Field diagnosis of chickpea diseases and their control. In: Information bulletin no. 28 (revised): International Crops Research Institute for the Semi-Arid Tropics. 2012.

Nicholson SE. A detailed look at the recent drought situation in the Greater Horn of Africa. J Arid Environ. 2014;103:71–9.

Pande S, Siddique K, Kishore G, Bayaa B, Gaur P, Gowda C, Bretag T, Crouch J. Ascochyta blight of chickpea (*Cicer arietinum L.*): a review of biology, pathogenicity, and disease management. Aust J Agric Res. 2005;56(4):317–32.

Pande S, Kishore GK, Upadhyaya H, Rao JN. Identification of sources of multiple disease resistance in mini-core collection of chickpea. Plant Dis. 2006;90(9):1214–8.

Pande S, Sharma M, Ghosh R, Rao S, Sharma R, Jha A. Opportunities for chickpea production in rain fed rice fallows of India. Baseline Survey Report. 2012.

Perttilä S, Valaja J, Jalava T. Apparent ileal digestibility of amino acids and metabolisable energy value in grains for broilers. Agric Food Sci. 2005;14(4):325–34.

Purushothaman R, Upadhyaya H, Gaur P, Gowda C, Krishnamurthy L. Kabuli and desi chickpeas differ in their requirement for reproductive duration. Field Crop Res. 2014;163:24–31.

Rasool S, Abdel Latef A, Ahmad P. Chickpea: role and responses under abiotic and biotic stress. In: Legumes under environmental stress: yield, improvement and adaptations. Chichester: Wiley; 2015. p. 67–79.

Rimal NS, Kumar S, Chahal V, Singh V. Impact of adoption of improved varieties of chickpea (*Cicer arietinum*) on yield and income in Madhya Pradesh. Indian J Agric Sci. 2015;85(4):555–60.

Rousta I, Javadizadeh F, Dargahian F, Ólafsson H, Shiri-Karimvandi A, Vahedinejad SH, Doostkamian M, Monroy Vargas ER, Asadolahi A. Investigation of vorticity during prevalent winter precipitation in Iran. Adv Meteorol. 2018;2018:1–13.

Saccardo F, Calcagno F. Consideration of chickpea plant ideotypes for spring and winter sowing. Options Méditerr. 1990;9:35–41.

Sagi MS, Deokar AA, Tar'an B. Genetic analysis of NBS-LRR gene family in chickpea and their expression profiles in response to Ascochyta blight infection. Front Plant Sci. 2017;8(838):1–14.

Salam MU, Davidson JA, Thomas GJ, Ford R, Jones RA, Lind beck KD, MacLeod WJ, Kimber RB, Galloway J, Mantri N. Advances in winter pulse pathology research in Australia. Australas Plant Pathol. 2011;40:549–67.

Santra D, Singh G, Kaiser W, Gupta V, Ranjekar P, Muehlbauer F. Molecular analysis of *Ascochyta rabiei* (Pass.) Labr, the pathogen of Ascochyta blight in chickpea. Theor Appl Genet. 2001;102(5):676–82.

Sarmah B, Acharjee S, Sharma H. Chickpea: crop improvement under changing environment conditions. In: Improving crop productivity in sustainable agriculture. Weinheim: Wiley-VCH; 2012. p. 361–80.

Saxena N, Saxena M, Johansen C, Virmani S, Harris H. Adaptation of chickpea in the West Asia and North Africa Region. Andhra Prasdesh: ICRISAT-ICARDA; 1996.

Sehulster L, Chinn R. Centres for Disease Control and Prevention Healthcare Infection Control Practices Advisory Committee. Guidelines for environmental infection control in health-care facilities. MMWR. 2003;52(RR10):1–42.

Serraj R, Bidinger F, Chauhan Y, Seetharama N, Nigam S, Saxena N. Management of drought in ICRISAT cereal and legume mandate crops. In: Water productivity in agriculture: limits and opportunities for improvement. Wallingford: CABI; 2003. p. 127–44.

Shafique A, Rehman S, Khan A, Kazi AG. Improvement of legume crop production under environmental stresses through biotechnological intervention. In: Emerging technologies and management of crop stress tolerance. Amsterdam: Elsevier; 2014. p. 1–22.

Shamsi S, Khatun A. Prevalence of fungi in different varieties of chickpea (*Cicer arietinum* L.) seeds in storage. J Bangladesh Acad Sci. 2016;40(1):37–44.

Sharma M, Ghosh R. An update on genetic resistance of chickpea to Ascochyta blight. J Agron. 2016;6(1):18.

Singh K. Winter chickpea: problems and potential in the Mediterranean region. Options Méditerr. 1990;9:25–34.

Singh K, Reddy M. Improving chickpea yield by incorporating resistance to Ascochyta blight. Theor Appl Genet. 1996;92(5):509–15.

Smith RA, Schuetz M, Karlen SD, Bird D, Tokunaga N, Sato Y, Mansfield SD, Ralph J, Samuels AL. Defining the diverse cell populations contributing to lignification in Arabidopsis stems. Plant Physiol. 2017;174(2):1028–36.

Sotelo A, Flores F, Hernández M. Chemical composition and nutritional value of Mexican varieties of chickpea (*Cicer arietinum* L.). Plant Food Hum Nutr. 1987;37(4):299–306.

Stagnari F, Maggio A, Galieni A, Pisante M. Multiple benefits of legumes for agriculture sustainability: an overview. Chem Biol Technol Agric. 2017;4(2):1–13.

Stevens DA. Diagnosis of fungal infections: current status. J Antimicrob Chemother. 2002;49(suppl_1):11–9.

Sujak A, Kotlarz A, Strobel W. Compositional and nutritional evaluation of several lupin seeds. Food Chem. 2006;98(4):711–9.

Summerfield R, Roberts E, Erskine W, Ellis R. Effects of temperature and photoperiod on flowering in lentils (*Lens culinaris* Medic.). Ann Bot. 1985;56(5):659–71.

Tanno K-I, Willcox G. The origins of cultivation of *Cicer arietinum* L. and *Viciafaba* L.: early finds from Tell el-Kerkh, North-West Syria, late 10th millennium Bp. Veg Hist Archaeobot. 2006;15(3):197–204.

Ten Have A, Espino JJ, Dekkers E, Van Sluyter SC, Brito N, Kay J, González C, van Kan JA. The *Botrytis cinerea* aspartic proteinase family. Fungal Genet Biol. 2010;47(1):53–65.

Tivoli B, Baranger A, Avila CM, Banniza S, Barbetti M, Chen W, Davidson J, Lindeck K, Kharrat M, Rubiales D. Screening techniques and sources of resistance to foliar diseases caused by major necrotrophic fungi in grain legumes. Euphytica. 2006;147(1–2):223–53.

Toker C, Lluch C, Tejera N, Serraj R, Siddique K. Abiotic stresses. In: Chickpea breeding and management. Wallingford: CABI; 2007. p. 474.

Trumbo P, Schlicker S, Yates AA, Poos M. Dietary reference intakes for energy, carbohydrate, fiber, fat, fatty acids, cholesterol, protein and amino acids. J Acad Nutr Diet. 2002;102(11):1621–30.

Uchida K, Akashi T, Aoki T. The missing link in leguminous pterocarpan biosynthesis is a dirigent domain-containing protein with isoflavanol dehydratase activity. Plant Cell Physiol. 2017;58(2):398–408.

Udall JA, Wendel JF. Polyploidy and crop improvement. Crop Sci. 2006;46(Supplement_1):S3–14.

Upadhyaya HD. Geographical patterns of variation for morphological and agronomic characteristics in the chickpea germplasm collection. Euphytica. 2003;132(3):343–52.

Van den Brink J, de Vries RP. Fungal enzyme sets for plant polysaccharide degradation. Appl Microbiol Biotechnol. 2011;91(6):1477–92.

Van der Maesen L. A monograph of the genus, with special references to the chickpea (*Cicer arietinum* L.) its ecology and cultivation. Wageningen: H Veenman & Zonen; 1972. p. 1–341.
Van der Maessen L. Cicer L.: a monograph of the genus, with special reference to the chickpea (*Cicer arietinum* L.), its ecology and cultivation. Wageningen: Veenman; 1972.
Venn B, Mann J. Cereal grains, legumes and diabetes. Eur J Clin Nutr. 2004;58(11):1443–61.
Vidal T, Lusley P, Leconte M, de Vallavieille-Pope C, Huber L, Saint-Jean S. Cultivar architecture modulates spore dispersal by rain splash: a new perspective to reduce disease progression in cultivar mixtures. PLoS One. 2017;12(11):e0187788.
Waliyar F, Kumar KVK, Diallo M, Traore A, Mangala U, Upadhyaya H, Sudin H. Resistance to pre-harvest aflatoxin contamination in ICRISAT's groundnut mini core collection. Eur J Plant Pathol. 2016;145(4):901–13.
Who J, Consultation FE. Diet, nutrition and the prevention of chronic diseases. World Health Organ Tech Rep Ser. 2003;916(i–viii):1–149.
Williamson B, Tudzynski B, Tudzynski P, van Kan JA. *Botrytis cinerea*: the cause of grey mould disease. Mol Plant Pathol. 2007;8(5):561–80.
Wilson C, Solar J, El Ghaouth A, Wisniewski M. Rapid evaluation of plant extracts and essential oils for antifungal activity against *Botrytis cinerea*. Plant Dis. 1997;81(2):204–10.

Chapter 11
Management of *Fusarium udum* Causing Wilt of Pigeon Pea

Gagan Kumar, Raina Bajpai, Basavaraj Teli, Jhumishree Meher, Md. Mahtab Rashid, and Birinchi Kumar Sarma

11.1 Introduction

Pigeon pea is an important source of protein and vitamin, and it is the second most edible legume crop after chickpea and contributes about 90% production of the total world production in India (Allen and Lenné 1998; Dhanasekar et al. 2010). Its protein and essential amino acid content makes it an important food in a vegetarian diet, with its seed and pod husk being the sources of feed (Varshney et al. 2010). In addition to protein and amino acid, it also contains carbohydrates, minerals, and fibers. Its plantation covered 4.3 million hectares globally (Anonymous 2007). In India pigeon pea production and productivity are 2.76 metric tons and 762 kg/ha, respectively, coming from an area of about 3.63 million hectare (the Year 2010, ICAR Vision 2030/2010). Thirty-two species belong to the genus *Cajanus*, and most of them are found in India and Australia, whereas only one species is native from West Africa. Pigeon pea can be grown under drought conditions with significant return and minimum input. In India pigeon pea productivity is low due to the lack of new cultivars and infection by plant pathogens (Nene et al. 1996). It is cultivated with a minimum input of fertilizers and disease management strategies. Pigeon pea production is affected by many biotic and abiotic stresses. Under biotic stress, several pathogens such as fungi bacteria, viruses, nematodes, and

G. Kumar (✉)
Department of Mycology and Plant Pathology, Institute of Agricultural Sciences, Banaras Hindu University, Varanasi, Uttar Pradesh, India

Department of Plant Pathology, School of Agriculture, Lovely Professional University, Phagwara, Punjab, India

R. Bajpai · B. Teli · J. Meher · M. M. Rashid · B. K. Sarma
Department of Mycology and Plant Pathology, Institute of Agricultural Sciences, Banaras Hindu University, Varanasi, Uttar Pradesh, India

B. P. Singh et al. (eds.), *Management of Fungal Pathogens in Pulses*, Fungal Biology, https://doi.org/10.1007/978-3-030-35947-8_11

mycoplasma-like organisms are responsible for the lower yield of pigeon pea (Nene et al. 1989; Kannaiyan et al. 1984). Some important diseases responsible for legume crop loss include *Fusarium* wilt, sterility mosaic, phytophthora blight, macrophomina root rot, alternaria leaf spot, and cercospora leaf spot caused by *Fusarium udum*, viruses, *Phytophthora drechsleri* f. sp. *cajani*, *Macrophomina phaseolina*, *Alternaria tenuissima*, and *Cercospora cajani*, respectively (Kannaiyan et al. 1984). These diseases and other abiotic factors such as low moisture stress, waterlogging, and salt stress are responsible for a significant reduction in yield of pigeon pea (Varshney et al. 2007; Saxena 2008). The diseases *Fusarium* wilt and sterility mosaic are economically important in our country. *Fusarium* wilt is a very severe disease, causing yield loss of about US $71 million annually in India. Wilt is a soilborne disease that affects the yield of crop significantly especially in wilt-susceptible cultivars (Reddy et al. 1990). *Fusarium udum* is soil inhabitant in nature and enters the vascular system of the plant through the root system. Because of the soilborne nature of wilt disease, management through cultural practices is very difficult at a significant level. Some chemical fungicides are effectively managing this disease, but the extreme use of chemicals is harmful and noneconomical. Biocontrol strategies are also in use through several antagonistic microorganisms for managing this disease (Chaudhary and Kumar 1999). Many fungal and bacterial commercial products are also developed for soilborne pathogen management (Kumar and Sarma 2016; Kumar et al. 2017). Use of these biocontrol antagonistic microorganisms and their commercial product in plant disease management is economical and risk-free concerning health hazards. In this chapter, we have discussed all the management strategies from conventional to advanced molecular technologies for wilt disease of pigeon pea.

11.2 History

In 1809, Link was the first scientist to narrate about the genus *Fusarium* – the pathogen with fusiform, nonseptate spores borne on a stroma. Later, a detailed account of *Fusarium* species and pigeon pea wilt was first reported by Butler (1906). In India, this destructive fungus was first described in 1906 by E.J. Butler in the pigeon pea crop from Bihar and hence named as *Fusarium udum* Butler and later reported in several other countries in Africa, South Asia, and Europe (Karimi et al. 2012). Then, *F. udum* was established as a new species by Butler (1910), and isolation and identification of the fungus were carried out. Previously, *F. oxysporum* f. sp. *udum* was used frequently. Extensive characterization of *Fusarium*-plant interaction in the prospect of its biochemistry and physiology has been already done; however, recognition of vital molecules involved in the pathogenesis of *Fusarium* sp. did not start till convenient molecular genetic techniques for filamentous fungi were available (Timberlake and Marshall 1989; Datta and Lal 2013). Due to the soilborne nature of the pathogen, chemical control is ineffective in many established cases, and managing the disease seems to be very challenging. However, deployment of resistant varieties is unlikely because of its high degree of genetic variability among the pathogenic population (Kumar and Upadhyay 2014). At the present scenario, three fungicides commonly used for the management of *Fusarium* wilt are thiram, benomyl, and bavistin (Vidhyasekaran et al. 1997; Meena et al. 2002; Melent'ev et al. 2006). Moreover, microorganisms producing various types of mycolytic enzymes (chitinases, glucanase, and proteases) have shown a substantial impact on disease development as they can degrade chitin and glucan present in the fungal cell wall (Deshpande 1999; Hillocks et al. 2000; Hoster et al. 2005; Patel et al. 2007).

11.3 Distribution

Worldwide, pigeon pea wilt causes considerable devastation to the production of pigeon pea (Kannaiyan et al. 1984). At crop blooming and maturity stages, 30–60% of disease incidence has been recorded; on the other hand, yield losses may increase up to 100% when susceptible cultivars were used (Okiror 2002; Dhar et al. 2005). It is extensively occurring in India, Malawi, and East Africa leading to more than 50% yield losses, and despite these, countries like Indonesia, Mauritius, Bangladesh, Grenada, Myanmar, Venezuela, Trinidad, Nevis, Nepal, and Tobago are well-known for incidence of *Fusarium udum* (Reddy et al. 2012; Marley and Hillocks 1996). In the Indian context, this disease was reported in most of the pigeon pea-growing states and caused about US$ 71 million annual production losses (Reddy et al. 2012) except in southern states. However, the heavy incidence was reported in Vidharbha (13.66%) followed by the Marathwada region where maximum severity recorded up to 90% in the state of Maharashtra (Shinde et al. 2014). In other states like Bihar, Jharkhand, Orissa, and West Bengal, *Fusarium* wilt was effectively found with a substantial range of cultural, morphological, and pathogenic variability in maximum isolates collected from pigeon pea-growing regions (Kumar and Upadhyay 2014). Mesapogu et al. (2012) have reported genetic diversity and pathogenic variability among 30 isolates of *Fusarium udum* collected from diverse agroclimatic conditions representing 7 states of India, i.e., Andhra Pradesh, Uttar Pradesh, Jharkhand, West Bengal, Haryana, Rajasthan, and Punjab.

11.4 Symptoms

The disease can be diagnosed by visualizing the gradual or sudden wilting of the pigeon pea plant. Similarly, the leaves show interveinal clearing followed by withering, yellowing, and drying of young leaves on the upper portion of the plant. Wilted plant loss their tugidity because off chlorosis and necrosis resulting in premature leaf drop and drooping of apical shoot followed by drying of entire shoot (Upadhyay and Rai 1992). As the pathogen survives in the soil and the nature of the infection is soilborne, it will infect the tap root system of pigeon pea plants resulting in wilting of the whole plant instead of partial wilting. If the stem of infected plants is split open, browning of vascular tissue mainly the xylem is the most common visible symptom which differentiates it from other diseases. The wilting symptoms are the most common and prominent during the flowering and pod maturation stages (Reddy et al. 1990). Another visible symptom is purple banding, which extends upward from the base of the plants and is easily seen on the stem portion. Purple banding helps in differentiating healthy and infected plants (Sharma et al. 2016).

11.5 Disease Development and Pathogenicity

Fusarium wilt of pigeon pea is both a soil-borne and seed-borne disease in which the infection level of untreated seeds may range from 13% to 19% (Kannaiyan et al. 1984). The infected seeds thus serve as a primary vehicle for the spread of this disease over long distances and/or to the newer areas. The pathogen, *Fusarium udum*, survives in the soil for more than 3 years on the infected plant detritus. The disease incidence and disease severity are principally dependent on the conditions of soil and the genotype of the crop. The incidence of disease in susceptible cultivars is facilitated by a slightly acidic to slightly alkaline soil having sand particles more than half percentage in their soil texture (Singh and Hussain 1964; Upadhyay 1979). A soil temperature of about 20–29 °C and soil moisture of about 6–16% are most suitable for the development of wilt disease in pigeon pea (Upadhyay 1979). As per the reports, disease incidence among different soils depends chiefly on the survival and saprophytic activity of the pathogen in those soils that are ultimately favored by the availability of the host substrate. The severity of the disease is dependent on the duration of the pigeon pea varieties as very short-duration varieties suffer less than the long-duration and medium-duration varieties. Growing of susceptible pigeon pea varieties over the infested soils repeatedly increases the disease severity and disease incidence.

Earlier the wilt of pigeon pea was known to be caused only by the imperfect state of the pathogen (*Fusarium udum*), but the discovery of its perfect state, i.e., *Gibberella indica* (Upadhayay and Rai 1983), is known to occur through both the stages. As the perfect state is not known to be present frequently under natural conditions, the imperfect state is most common to incur the disease. In both the states, the pathogen is known to grow externally and internally through the production of a mycelial mass and conidia on the host's surface, majorly on the collar region and roots (Upadhyay and Rai 1982). After the surface colonization, the fungal hyphae invade the fine branches of roots that grow laterally and continue to proliferate in the vessels of xylem. Even though the infection may take place in the seedling stage of the plant, but the expression of disease is maximum during flowering and the podding stage of plants (Reddy et al. 1998), which can be due to the longer time required by the pathogen for colonization in the plants. It takes approximately about 3–4 months for the fungus to cause wilting in the infected plants which are when the basal half of the main stem is colonized by the pathogen (Reddy et al. 1998). This is the reason that can be understood as to why the short-duration crops have low levels of wilt infestation when compared to long-duration crops as the former ones are escaping the wilt incidence.

Once the infected plants wilt and die, the pathogen continues to live and survive as a saprophyte for many years, mainly on the dead plant parts in its perfect form (Upadhayay and Rai 1983) or imperfect form (Nene et al. 1980). Both the states of the fungus survive simultaneously on the host plant. In addition to the confinement of pathogen survival mainly on the dead roots and debris of infected plants, it may survive on the other organic matter for a limited period. Apart from these, the fungus *Fusarium udum* also survives on other fungi in the soil as mycoparasite as well as on the bodies of termites that feed on the wilted host roots (Upadhyay and Rai 1982, 1983). The chlamydospores are also known to be formed in both the phases of the fungus, i.e., the parasitic and the saprophytic phases, depending on the environmental conditions from the hypha and the conidia (Sinha 1975). The fungus has been also observed to produce a large number of dark violet perithecia on the exposed roots and collar region of the host plant which also serves as resting structures. These *Fusarium udum* perithecia produce ascospores in large numbers which remain physiologically inactive in the soil for a limited period and after which they produce either conidia or somatic hyphae on germination leading to infection of the pigeon pea plants (Rai and Upadhyay 1982).

In recent years, many of the studies on morphological, cultural characterization and the rate of reaction of the pathogen *Fusarium udum* have provided enough evidence for the existence of different virulence groups (Harlapur et al. 2007; Mahesh et al. 2010; Karimi et al. 2010). The variable reactions of various tested resistant pigeon pea varieties show the possibility of the presence of different physiological forms of the pathogen (Muhammad et al. 2011). In a study, Reddy et al. (1998) reported three strains of the pathogen which showed sensitivity/or resistance against several pigeon pea differentials.

11.6 Mechanism of Host Plant Resistance

The employment and use of resistant varieties of the crop is the most economical, effective, and eco-friendly strategy for the control of diseases even though their response to the cultivating conditions will be a subject of concern (Saxena et al. 2012). To come up with a sound breeding program for the development of disease-resistant crop varieties, we need to understand the mechanism of host plant resistance and what mechanism to strengthen up in plants to restrict pathogen invasion. There are mainly two mechanisms that constitute host plant resistance, viz., constitutive and induced defense mechanisms. The constitutive resistance mechanisms contain all the preformed chemical factors and physical barriers that are present in the host plant in advance to the attack of phytopathogens (Dangl and Jones 2001). The physical barriers consist of the thick and/or hard cuticle, wax deposition in the epidermal cells, stomatal shape and size, and the pericycle of the root (Keen 1992).

The chemical factors of the constitutive defense mechanism consist of peptides, proteins, protein inhibitors, preformed secondary metabolites, alkaloids, phenols, phytoanticipins, etc., which add up to the early barriers of defense being a part of plant's natural growth and development (Heath 2000; Dixon 2001; Grayer and Kokubun 2001). The plants are also reported to exudate some fungi toxic substances that restrict and/or inhibit the spore germination of the phytopathogen (Agrios 2004).

The induced defense mechanisms are the ones which get triggered on after the attack of phytopathogen and involve both chemical and physical factors (Agrios 2004). The most important step of induced defense mechanism is the recognition of the phytopathogen by the host plant so that it can conjure the defense reactions (Dixon et al. 1994; Schenk et al. 2000). The process of reaction starts with the recognition of the molecular pattern of the pathogen and is termed as pathogen-associated molecular patterns (PAMP) (Nürnberger and Lipka 2005). This recognition of the pathogen leads to signal transduction involving a cascade of biochemical events which leads to incitation of defense responses (Keen 1992; Dixon et al. 1994; Baron and Zambryski 1995). The most frequent defense response is the hypersensitive response (De Wit 1992) which is a form of programmed cell death (Greenberg and Yao 2004). The hypersensitive reaction restricts the growth of the fungus to newer plant cells (Tomiyama 1982; Keen 1992; Schenk et al. 2000). In addition to this, the other induced reactions include rapid oxidative burst, ion fluxes, and strengthening of the cell wall by increased synthesis of cellulose, lignin, phenolic compounds, and hydroxyproline-rich glycoproteins (Bowels 1990; Agrios 2004). The rapid oxidative burst is mainly through the production of hydroxyl radical (OH), hydrogen peroxide (H_2O_2), and superoxide (O_2^-), and these reactive oxygen species impart cross-linkage of the proteins present in the cell wall of the plant resistant to fungal enzyme attack (Bradley et al. 1992; Keen 1999). These reactive oxygen species are also known to induce hypersensitive cell death while working as an agent in the cell signaling process (Levine et al. 1994; Alvarez et al. 1998).

There are other defense mechanisms which constitute in host plant resistance, and it comprises of production of vascular occlusions such as tyloses and gels (Mace 1963) and defense-related gene expression involving the production of suberin and lignin, signal transduction proteins, phytoalexins, and pathogenesis-related proteins (Reymond and Farmer 1998; Greenberg and Yao 2004). The production of the signaling compounds in the host plant after the recognition of the phytopathogen attack leads to the enactment of defense reactions systemically throughout the plant and is termed systemic resistance (Ryals et al. 1994).

11.7 Management of Fusarium Wilt Disease

There are different methods for the control and management of *Fusarium udum* followed in agricultural technology with its positive and negative impacts. For complete resistance, single, race-specific resistance genes (R genes) could be used. For incomplete resistance, a bunch of minor genes work together for broad-spectrum effect. Complete management of fungal disease is difficult due to lack of knowledge regarding plant-pathogen interaction at genetic, histological, and molecular levels. Thus, to protect pigeon pea from *Fusarium* in a sustainable way, it is necessary to build a novel and potential approach by investigating the existing technologies. Some of the important control methods are discussed here.

11.7.1 Cultural Management

For the formation of barrier in pigeon pea against *Fusarium* wilt, numerous cultural practices are used. Among them, crop rotation is one of the best control measures. Crops like tobacco (*Nicotiana tabacum* L.), sorghum (*Sorghum bicolor* (L.) Moench), or castor (*Ricinus communis* L.) are rotated with pigeon pea for 3 years to wipe out the pathogen completely from the field. To reduce the infestation percentage below 20%, cultivation of the main crop could be followed with a year break with sorghum, or the land could be left fallow. The application of farmyard manure or *Crotalaria juncea* as green manure also reduces the incidence of wilt to a significant level (Ingole et al. 2005). Another method is reducing *Fusarium* inoculums from the field by solarization technique during the summer season (Reddy et al. 2012). Intercropping of sorghum with pigeon pea reduces incidences to 24% as compared to the sole crop which gets 85% incidence (Natarajan et al. 1985). Mixed cropping of *Crotalaria medicaginea* also has a positive impact on reducing wilt (Upadhyay and Rai 1981).

11.7.2 Chemical Management

Chemical management is one of the most effective and common measures. An equivalent mixture of benomyl and thiram is used for seed treatment and considered effective (Reddy et al. 2012). Use of biocontrol agent like formulation of *Trichoderma viride* and farmyard manure (2 kg and 125 kg, respectively) for one square measure is also found to be very successful in reducing *Fusarium* wilt (Perchedpied and Pitrat 2004). Addition of mineral in the soil like boron (Bo), zinc (Zn), manganese (Mn), and methyl bromide (CH_3Br) diminishes the disease event of *Fusarium* wilt (Maisuria et al. 2008). For effective management of this disease, antibiotics like bulbiformin and griseofulvin have also been accounted.

11.7.3 Biological Management

As chemicals lead to undesirable and harmful effects on various living entities, moreover it also causes an imbalance in the ecosystem. Thus, it creates a need for a healthy control measure. The use of biological agents is thus a significant measure as it is a member of the ecosystem and a potential antagonist to pathogens. According to a few reports, addition of antagonists in the soil diminishes the *Fusarium udum* incidence (Maisuria et al. 2008; Bapat and Shar 2000; Singh et al. 2002; Anjaiah et al. 2003). Various rhizobacteria as biocontrol agents are used for its management (Siddiqui 2006; Siddiqui and Shakeel 2007; Pusey 1989; Bapat and Shar 2000; Siddiqui et al. 2005). The addition of *T. harzianum* provides disease control of 22–61.5% at all pathogen levels (Prasad et al. 2002). According to reports population of *F. udum* is drastically reduced by antagonism of *Aspergillus terreus*, *Aspergillus niger*, *Micromonospora globosa*, and *Aspergillus flavus* (Upadhyay and Rai 1981) in a biocontrol experiment. In naturally infested soil, the addition of *Pseudomonas aeruginosa* PAN1 significantly suppresses the incidence of *Fusarium* in pigeon pea and chickpea (Anjaiah et al. 2003). A graphical representation of direct and indirect mechanisms of biocontrol is presented in Fig. 11.1.

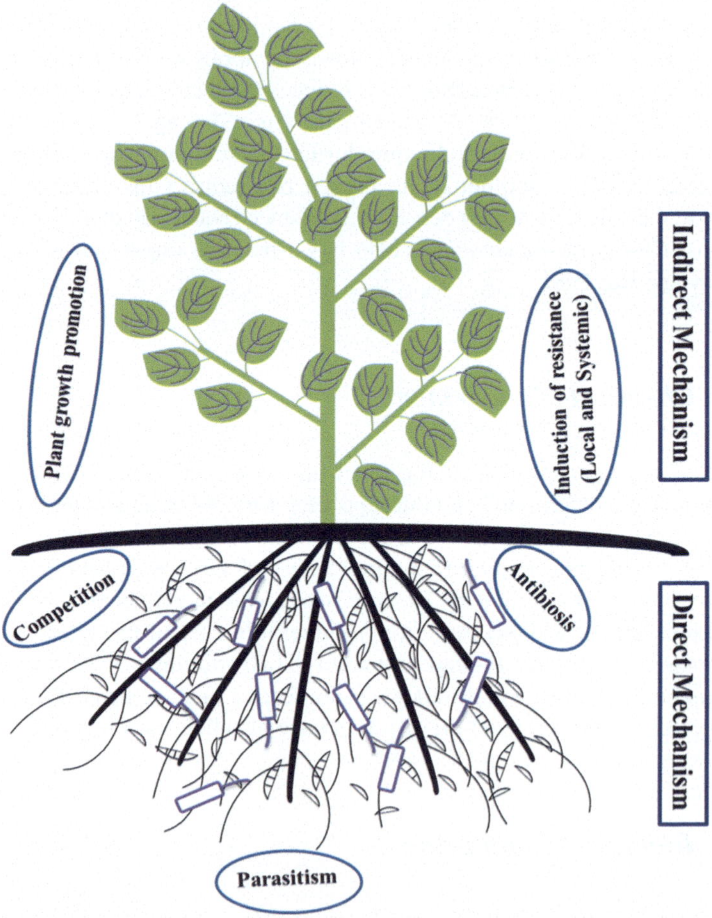

Fig. 11.1 Diagram represents the mechanisms of biocontrol agent used for disease management

11.7.4 Transcriptomics Approaches

Plant receptor protein recognizes the pathogen-derived molecule which is the initial step in defense response by activation of signal transduction cascades which triggers expression of various plant defense genes (Barilli et al. 2014). The study of gene expression provides a detailed knowledge regarding genes which were differentially expressed and various metabolic conduits at the time of host-pathogen interfaces. It can jointly help to unveil candidate resistant genes collaborating in every step of plant defense response (Ichinose et al. 2001). In the era of molecular plant breeding, marker-assisted selection (MAS) could be highly useful by applying the knowledge of the defense-responsive genes in legumes against fungal pathogen attack to legume plants, and under transformation event, any change in expression of such candidate genes could be linked with improved resistance. There are certain techniques used in transcriptomics like enhancing the potential number of defense-related genes by generating cDNA (complementary DNA) libraries from plants under stress against pathogens inoculation or elicitor-treated tissues or cells. The second one is the application of macro- or microarray designed by using orthologue sequences from other legumes in the format of unigenes, cDNA, expressed sequence tags (ESTs), or resistance gene analogs (RGAs) in the query legumes like pigeon pea under specific fungal stress conditions. These methods help to identify transcripts that are induced under pathogenic attacks and majorly associated with candidate resistant genes with a certain level of expression. Transcriptomics also helps to explore the information of genome sequence information with the aid of new less expensive sequencing platforms (Illumina (Solexa) sequencing, Roche 454 sequencing, Ion Torrent (Proton/PGM sequencing), and SOLiD sequencing). NGS technologies decrease the complexity of transcriptome techniques like SSH, cDNA-AFLP, SuperSAGE (serial analysis of gene expression), or MPSS (massive parallel signature sequencing), thereby increasing the identified transcript amount devoid of cloning and Sanger sequencing. Now, RNAseq technique allows building de novo transcriptomics that generates the transition of the transcript in expression form of both plant host and the inoculated fungal pathogen for examining plant-pathogen interactions, in addition to its basic work of studying all expressed transcript's sequencing at that particular time (Tadege et al. 2009).With the help of transcriptome profiling techniques, numerous diverse expressed genes population across the genome can be easily generated under pathogen attack. It is difficult to differentiate such a transcript associated with defense response and resistant phenotypes. This can be resolved by studying their co-localization with quantitative trait loci (QTLs) and exploring their functional analysis. Different advanced molecular techniques like gene silencing via RNA interference (RNAi) and virus-induced gene silencing (VIGS) are also used nowadays for knowing functional activities of PR proteins and biotic stress-induced genes (Tadege et al. 2009). A generalized presentation of phases showing the involvement of transcription factor in the induction of systemic acquired resistance against pathogen stress is presented in Fig. 11.2.

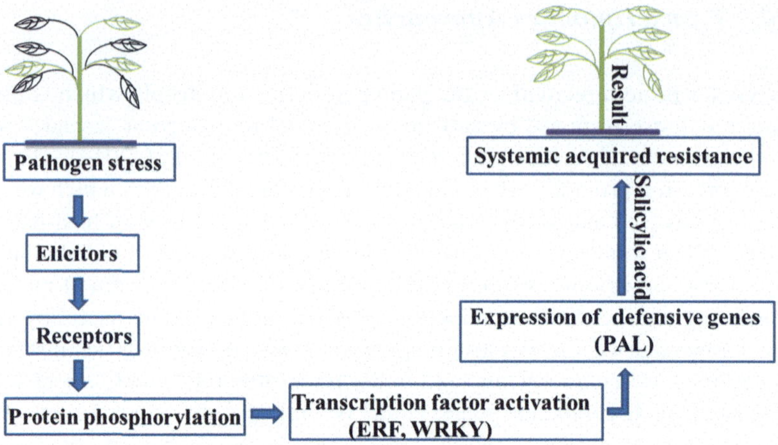

Fig. 11.2 A generalized presentation of phases showing involvement of transcription factor in induction of systemic acquired resistance against pathogen stress. Protein phosphorylation occurs early with the recognition of pathogen elicitor by host receptor. Further transcription factor activation induces expression of defense genes such as PAL. Salicylic acid biosynthesis and defense gene activate systemic acquired resistance during plant pathogen interaction

11.7.5 Proteomics Approaches

Protein expression and its functional activity rely on the extent of expression of genes and posttranscriptional and posttranslational regulations. Therefore there could be a large chance that all transcripts derived from the successful expression of mRNA do not form successful protein accumulation and function. Thus, it is also significant to study protein accumulation to get a clear picture of the mechanisms of plant-pathogen interaction. Recent proteomic technologies provide opportunities for large-scale protein profiling via quantitative and qualitative methods (Qin et al. 2013). In comparative proteomics, protein is separated by electrophoresis based on their mass and isoelectric points followed by spectrometry techniques based on protein identification like de novo sequencing or peptide mass fingerprinting. Another technique is a separation of chromatography-based peptide mixtures continuing their detection through mass spectrometry (Nautrup-Pedersen et al. 2010) and shotgun proteomics which analyzes direct tandem mass spectrometric analysis that includes chromatographic separation based on cell lysis (Qin et al. 2013). All these techniques are practiced in legume particularly in the establishment of subcellular localization of target proteins, thus forming reference protein maps (Salavati et al. 2012). But, in legumes after pathogen attack, the study of proteomics is quiet far lacking behind as compared to other molecular advancements. But there is an example of a proteome study in chickpea – *Fusarium oxysporum* (Bourgeois et al. 2011). To detect protein variation under biotic stresses, comparative proteomic approaches are highly significant. Thus, there is a huge expectation from proteomic techniques that might unveil endogenous elements that provide resistance to fungal diseases.

11.8 Conclusion

The use of resistant variety is the most effective way to restrict the incidence of a disease. At present in the molecular biology and biotechnology era, it is possible to know about the genes, enzymes, proteins, and transcription factors that show a highly active defense response against pathogen attack. The study of resistances sources (Genes, protein etc.) can be beneficial for developing resistace in crop plant. For this purpose the current biotechnological and molecular biology techniques provide knowledge on transcription factors to detect stress-responsive genes of the plant. Further proteomics and genomics information is mandatory to know all cellular processes under stress response for better crop improvement.

References

Agrios GN. Plant pathology. 5th ed. London: Academic Press; 2004.

Allen DJ, Lenné JM. Pathology of food and pasture legumes. CAB International in association with the International Crops Research Institute for the Semi-Arid Tropics (ICRISAT); 1998.

Alvarez ME, Pennell RI, Maijer PJ, Ishikawa A, Dixon RA, Lamb C. Reactive oxygen intermediates mediate a systemic signal network in: the establishment of plant immunity. Cell. 1998;92:773–84.

Anjaiah V, Cornelis P, Koedam N. Effect of genotype and root colonization in biological control of Fusarium wilts in pigeonpea and chickpea by *Pseudomonas aeruginosa* PNA1. Can J Microbiol. 2003;49:85–91.

Bapat S, Shar AK. Biological control of Fusarial wilt of pigeon pea by *Bacillus brevis*. Can J Microbiol. 2000;46:125–32.

Barilli E, Rubiales D, Gjetting T. Differential gene transcript accumulation in peas in response to powdery mildew (*Erysiphe pisi*) attack. Euphytica. 2014;198:13–28.

Baron C, Zambryski PC. The plant response in pathogenesis, symbiosis, and wounding: a variation on a common theme. Annu Rev Genet. 1995;29:107–29.

Bourgeois M, Jacquin F, Cassecuelle F. A PQL (protein quantity loci) analysis of mature pea seed proteins identifies loci determining seed protein composition. Proteomics. 2011;1:1581–94.

Bowels DJ. Defense related proteins in higher plants. Annu Rev Biochem. 1990;59:873–907.

Bradley DJ, Kjellbom P, Lamb CJ. Elicitor- and wound – induced oxidative cross-linking of a proline-rich plant cell wall protein: a novel, rapid defense response. Cell. 1992;70:21–30.

Butler EJ. The wilt disease of Pigeonpea and Pepper. Agri J Ind. 1906;1:25–6.

Butler EJ. The wilt disease of Pigeonpea and parasitism of *Neocosmospora vasinfecta*. Depart Agri Bull Ind. 1910;2:1.

Chaudhary RG, Kumar K. Potentials of biocontrol agents against milt in pre rabi pigeonpea crop. Indian J Plant Pathol. 1999;17:67–9.

Dangl JL, Jones JDG. Plant pathogens and integrated defense responses to infection. Nature. 2001;41:26–833.

Datta J, Lal N. Genetic diversity of Fusarium wilt races of pigeonpea in major regions of India. Afr Crop Sci J. 2013;21(3):201–11.

Deshpande MV. Mycopesticide production by fermentation: potential and challenges. Crit Rev Microbiol. 1999;25:229–43.

De Wit PJGM. Molecular characterization of gene-for-gene systems in plant-fungus interactions and the application of avirulence genes in control of plant pathogens. Ann Rev Phytopathol 1992;30:391–418.

Dhanasekar P, Dhumal KN, Reddy KS. Identification of RAPD markers linked to the plant type gene in pigeonpea. Indian J Biotechnol. 2010;9:58–63.
Dhar V, Reddy MV, Chaudhary RG, Aliand M, Kumar S. Major diseases of pigeonpea and their management. IIPR Kanpur. 2005:229–61.
Dixon RA, Harrison MJ, Lamb CJ. Early events in the activation of plant defense responses. Annu Rev Phytopathol. 1994;32:479–501.
Dixon RA. Natural products and plant disease resistance. Nature. 2001;411(6839):843–7.
Grayer RJ, Kokubun T. Plant–fungal interactions: the search for phytoalexins and other antifungal compounds from higher plants. Phytochemistry. 2001;56(3):253–63.
Greenberg JT, Yao N. The role and regulation of programmed cell death in plant-pathogen interactions. Cell Microbiol. 2004;6:201–11.
Harlapur SI, Kulkarni MS, Yeshoda H, Srinkant K. Variability in *Exserohilum turcicum* (pass) Leonard and Suggs, causal agent of turcicum leaf blight of maize. J Agri Sci. 2007;20(3):665–6.
Heath MC. Hypersensitive response-related death. Plant Mol Biol. 2000;44:321–34.
Hillocks RJ, Minja E, Mwaga A. Diseases and pests of pigeonpea in eastern Africa: a review. Int J Pest Man. 2000;46:7–18.
Hoster F, Schmitz JE, Daniel R. Enrichment of chitinolytic microorganisms: isolation and characterization of a chitinase exhibiting antifungal activity against phytopathogenic fungi from a novel Streptomyces strain. Appl Microbiol Biotechnol. 2005;66:434–42.
Ichinose Y, Hisayasu Y, Sanematsu S. Molecular cloning and functional analysis of pea cDNA E86 encoding homologous protein to hypersensitivity-related. J Plant Sci. 2001;160:997–1006.
Ingole MN, Ghawade RS, Raut BT. Management of Pigeonpea wilt caused by *Fusarium udum* Butler. Crop Prot Prod. 2005;1:67–9.
Kannaiyan J, Nene YL, Reddy MV, Ryan JG, Raju TN. Prevalence of pigeonpea diseases and associated crop losses in Asia, Africa and America. Trop Pest Manage 1984;30:62–71
Karimi R, James OO, Silim SN. Inheritance of Fusarium wilt resistance in pigeonpea [*Cajanus cajan* (L.) Millspaugh]. Ind J Genet. 2010;70(3):271–6.
Karimi R, Owuoche JO, Silim SN. Importance and management of Fusarium wilt (*Fusarium udum* Butler) of pigeonpea. Intl J Agron Agric Res. 2012;2:1–14.
Keen NT. The molecular biology of disease resistance. Plant Mol Biol. 1992;19:109–22.
Keen NT. Plant disease resistance: Progress in basic understanding and practical application. Adv Bot Res. 1999;30:291–328.
Kumar G, Maharshi A, Patel JS, Mukherjee A, Singh HB and Sarma BK. Trichoderma : A Potential Fungal Antagonist to Control Plant Diseases. SATSA Mukhapatra – Annual Technical Issue. 2017;21:206–218.
Kumar S, Upadhyay J. Studies on cultural morphological and pathogenic variability in isolates of *Fusarium udum* causing wilt of pigeonpea. Indian Phytopathol. 2014;67:55–8.
Kumar G, Sarma BK. Ecofriendly management of soil-borne plant pathogens through plant growth promoting rhizobacteria. SATSA Mukhapatra - Annual Technical Issue. 2016;20:167–171.
Mace ME. Histochemical localization of phenols in healthy and diseased banana roots. Physiol Plant. 1963;16:915–25.
Mahesh M, Muhammad S, Prasad PS, Sreenivasa S. Studies on cultural variability of *Fusarium udum* isolates in India. Int J Sci Nat. 2010;1(2):219–25.
Maisuria VB, Gohel V, Mehta AN, Patel RR, Chhatpar HS. Biological control of Fusarium wilt of pigeonpea by *Pantoea dispersa*, a field assessment. Ann Microbiol 2008;58(3):411–419.
Marley PS, Hillocks RJ. Effect of root-knot nematodes (*Meloidogyne* spp.) on Fusarium wilt in pigeonpea (*Cajanus cajan*). Field Crop Res. 1996;46(1):15–20.
Meena B, Radhajeyalakshmi R, Marimuthu T. Biological control of groundnut late leaf spot and rust by seed and foliar application of powder formulation of *Pseudomonas fluorescens*. Biocontrol Sci Tech. 2002;12:195–204.
Melent'ev AI, Helisto P, Kuz'mina LY. Use of antagonistic bacilli for biocontrol of fungi degrading fresh wood. Appl Biochem Microbiol. 2006;42:62–6.
Mesapogu S, Bakshi A, Babu BK. Genetic diversity and pathogenic variability among Indian isolates of *Fusarium udum* infecting pigeonpea (*Cajanus cajan* (L.) millsp.). Int Res J Agri Sci Soil Sci. 2012;2(1):51–7.

Muhammad NS, Shahbaz TS, Safdar H, Anser A, Javaid I, Kiran H, et al. Evaluation of various fungicides for the control of gram wilt caused by *Fusarium oxysporium* f. sp. *ciceris*. Afri J Agri Res. 2011;6(19):4555–9.
Natarajan M, Kannaiyan J, Willey RW. Studies on the effects of cropping system on Fusarium wilt of pigeonpea. Field Crop Res. 1985;1:333–46.
Nautrup-Pedersen G, Dam S, Laursen BS. Proteome analysis of pod and seed development in the model legume *Lotus japonicus*. J Proteome Res. 2010;9:5715–26.
Nene YL, Kannaiayan J, Haware MP, Reddy MV. Review of the work done at ICRISAT on soil-borne diseases of pigeon pea and chickpea. In: Proceedings of the consultants group discussion on the resistance to soil borne diseases of Legumes, 8–11 January 1979, ICRISAT Center, Patancheru, AP India, 3–39; 1980.
Nene YL, Sheila VK, Sharma SB. A World list of chickpea and pigeonpea pathogens. Legume Pathol Prog Rep. 1989;7:23.
Nene YL, Sheila VK, Sharma SB. A world list of chickpeas and pigeonpea pathogens. 5th ed; 1996. International Crops Research for the Semi-Arid Tropics, Patancheru, Andhra Pradesh – 502 324, India.
Nürnberger T, Lipka V. Non-host resistance in plants: new insights into an old phenomenon. Mol Plant Pathol. 2005;6:335–45.
Okiror MA. Genetics of resistance to *Fusarium udum* in pigeonpea [*Cajanus cajan* (L.) Millsp]. Indian J Genet Plant Br. 2002;62(3):218–20.
Patel B, Gohel V, Raol B. Statistical optimization of medium components for chitinase production by *Paenibacillus sabina* strain JD2. Ann Microbiol. 2007;57:589–97.
Perchedpied L, Pitrat M. Polygenic inheritance of partial resistance to *Fusarium oxysporium* f. sp. *melonis* race 1.2 melon. Phytopathology. 2004;94:1331–6.
Prasad RD, Rangeshwaran R, Hegde SV. Effect of soil and seed application of *Trichoderma harzianum* on pigeonpea wilt caused by *Fusarium udum* under field conditions. Crop Prot. 2002;21:293–7.
Pusey PL. Use of *Bacillus subtilis* and related organisms as biofungicides. Pestic Sci. 1989;27:133–40.
Qin J, Gu F, Liu D. Proteomic analysis of elite soybean Jidou17 and its parents using iTRAQ-based quantitative approaches. Proteome Sci. 2013;11:12.
Rai B, Upadhyay RS. *Gibberella indica*: the perfect state of *Fusarium udum*. Mycologia. 1982;74(2):343–6.
Reddy MV, Nene YL, Kannaiyan J, Raju TN, Saka VN, Davor AT, et al. Pigeonpea lines resistant to wilt in Kenya and Malawi. Intl Pigeonpea News. 1990;6:34.
Reddy MV, Raju TN, Lenne JM. Diseases of pigeon pea. In: Allen DJ, Lenné JM, editors. The pathology of food and pasture legumes. Wallingford: CAB International; 1998. p. 517–58.
Reddy MV, Raju TN, Sharma SB. Handbook of pigeonpea diseases. Information bulletin, ICRISAT, Patancheru. 2012; 42:12.
Reymond P, Farmer EE. Jasmonate and salicylate as global signals for defense gene expression. Cur Opin Plant Biol. 1998;1:404–11.
Ryals JA, Uknes SJ, Ward ER. Systemic acquired resistance. Plant Physiol. 1994;104:1109–12.
Salavati A, Taleei A, Bushehri AA, Komatsu S. Analysis of the proteome of common bean (*Phaseolus vulgaris* L.) roots after inoculation with *Rhizobium etli*. Protein Peptide Lett. 2012;19:880–9.
Saxena KB. Genetic improvement of pigeonpea - a review. Trop Plant Biol. 2008;1:159–78.
Saxena KB, Kumar RV, Saxena RK, Sharma M, Srivastava RK, Sultana R, Varshney RK, Vales MI, Pande S. Identification of dominant and recessive genes for resistance to Fusarium wilts in pigeonpea and their implication in breeding hybrids. J Euphy. 2012;188(2):221–7.
Schenk PM, Kazan K, Wilson I, Anderson JP, Richmond T, Somerville SC, Manners JM. Coordinated plant defense responses in Arabidopsis revealed by microarray analysis. Proc Natl Acad Sci U S A. 2000;97:11655–60.
Sharma M, Ghosh R, Telangre R, Rathore A, Saifulla M, Mahalinga DM, Saxena DR, Jain YK. Environmental influences on pigeonpea-Fusarium udum interactions and stability of genotypes to Fusarium wilt. Front Plant Sci. 2016;7:253.

Shinde VS, Zagade SN, Chavan AA. Cultural and morphological variation in *Fusarium udum*. J Plant Dis Sci. 2014;9(2):237–44.

Siddiqui S, Siddiqui ZA, Iqbal A. Evaluation of fluorescent pseudomonads and Bacillus isolates for the biocontrol of wilt disease complex of pigeon pea. World J Micrbiol Biotechnol. 2005;21:729–32.

Siddiqui ZA. PGPR: Prospective biocontrol agents of plant pathogens. In: Siddiqui ZA, editor. Biocontrol and biofertilization: Springer; 2006. p. 111–42.

Siddiqui ZA, Shakeel U. Screening of Bacillus isolates for potential biocontrol of the wilt disease complex of pigeon pea (*Cajanus cajan*) under greenhouse and small-Scale field conditions. J Plant Pathol. 2007;89:179–83.

Singh GP, Hussain A. Presence of fusaric acid in wilt affected pigeonpea plants. Curr Sci. 1964;33:287.

Singh R, Singh BK, Upadhyaya SK. Biological control of Fusarium wilt disease of Pigeonpea. J Plant Pathol. 2002;18:279–83.

Sinha AK. Control of Fusarium wilt of pigeon pea with Bavisitin, a systemic fungicide. Curr Sci. 1975;44:700.

Tadege M, Wang TL, Wen J. Mutagenesis and beyond! Tools for understanding legume biology. Plant Physiol. 2009;151:978–84.

Timberlake WE, Marshall MA. Genetic engineering of filamentous fungi. Science. 1989;244:1313–7.

Tomiyama K. Hypersensitive cell death: its significance and physiology in plant infection. In: Asada Y, Busnell WR, Ouchi S, Vance CP, editors. The physiological and biochemical basis. Berlin: Springer; 1982.

Upadhayay RD, Rai B. A new disease cycle of wilt of pigeon-pea. Curr Sci. 1983;52(20):978–81.

Upadhyay RS. Ecological studies on *Fusarium udum* Butler causing wilt disease of pigeonpea. Ph. D. Thesis, Banaras Hindu University; 1979.

Upadhyay RS, Rai B. Effect of cultural practices and soil treatments on incidence of wilt disease of pigeonpea. Plant Soil. 1981;62:309–12.

Upadhyay RS, Rai B. Ecology of *Fusarium udum* causing wilt disease of pigeon pea: population dynamics in the root region. Trans Br Mycol Soc. 1982;78(2):209–20.

Upadhyay RS, Rai B. Wilt of pigeonpea. In: Singh AN, Mukhopadhyay J, Kumar J, Chaube HS, editors. Plant diseases of international importance, Cereals and Pulses, vol. I. New Jersey: Prentice Hall; 1992. p. 388–414.

Varshney RK, Hoisington DA, Upadhyaya HD, Gaur PM, Nigam SN, Saxena KB, et al. Molecular genetics and breeding of grain legume crops for the semi - arid tropics. In: Varshney RK, Tuberosa R, editors. Genomic– Assisted crop improvement. The Netherlands: Springer; 2007. p. 207–41.

Varshney RK, Thudi M, May GD, Jackson SA. Legume genomics and breeding. Plant Breed Rev. 2010;33:257–304.

Vidhyasekaran P, Sethuraman K, Rajappan K. Powder formulations of *Pseudomonas fluorescens* to control pigeonpea wilt. Biol Control. 1997;8:166–71.

Chapter 12
Role of Biofertilizer in Biological Management of Fungal Diseases of Pigeon Pea [(*Cajanus cajan*) (L.) Millsp.]

Surbhi Gupta, Nidhi Didwania, and N. Srinivasa

12.1 Introduction

The world population is increasing at a high growth rate and is expected to reach ~9.6 billion in 2050 according to a recent United Nations report (UNPAN 2010). With a projected emphasis on sustainable genetic improvement of major staple crops including rice, wheat and maize, it is also important to lay light on the production of protein-rich foods to reduce global malnutrition and hunger. Proteins are the foremost building block of the human system. It is a known fact that developing countries have only 33% of the normal requirement of protein, hence making it a challenge for various nutritional development programs to fulfil the protein demand.

Leguminous plants (legumes or pulses) are one of the best available protein sources that can contribute a handful amount of proteins in the diet of developing countries as they require minimum care during cultivation and low inputs. Pigeon pea or red gram (*Cajanus cajan* (L.) Millsp.) occupies a chief place in worldwide agriculture among different legume crops (Saxena et al. 2010). It occupies 5.4 million hectares in 22 countries in the continents of Asia and Africa. Out of this India alone has more than 3.9 million hectares, i.e. 72% of the area, of all the pigeon pea-growing countries of the world (FAOSTAT 2018). Uttar Pradesh is the largest producer of pigeon pea in India, but the average yield released by the crop is much less than its other neighbouring states like Bihar and Jharkhand (Ahlawat et al. 2005; Prasad et al. 2017).

Pigeon pea (*Cajanus cajan* (L.) Millsp.) is the most vital legume crop in the world. India is one of the largest producers of pigeon pea commonly known as "arhar" in its northern part followed by the eastern side of Africa and Central

S. Gupta · N. Didwania (✉)
Manav Rachna International University, Faridabad, Haryana, India

N. Srinivasa
Department of Plant Pathology, Indian Agricultural Research Institute, New Delhi, India

B. P. Singh et al. (eds.), *Management of Fungal Pathogens in Pulses*, Fungal Biology, https://doi.org/10.1007/978-3-030-35947-8_12

America. It is roughly cultivated in at least 25 tropical and sub-tropical countries. This crop is greatly influenced by weather conditions; it is well raised in semi-arid tropical areas which are rain-fed. Cropping of pigeon pea is intermixed with maize, sorghum, pearl millet and some other legume crops like groundnut etc. It supplements soil through nitrogen fixation.

The term "biofertilizers" refers to live microbial culture, which when applied to plants, soil or composting pits helps in mobilization of various nutrients by their biological activity. Application of biofertilizers such as plant growth-promoting rhizobacteria (PGPR) and arbuscular mycorrhizal fungi (AMF) in agricultural field soils is well known. Assessment of native microbial field community is indispensable for developing tracing tools to monitor the introduced biofertilizers. Pigeon pea is affected by almost 60 plant pathogens comprising of bacteria, nematodes, fungi, viruses, etc., but luckily, only a few of them are of economic importance. Out of which, it is withered by numerous fungal diseases, viz. fusarium wilt, Phytophthora blight, Phoma stem canker, Alternaria blight and Macrophomina root rot.

12.2 Some Major Fungal Diseases of Pigeon Pea

Diseases of economic importance in the country are fusarium wilt caused by *Fusarium udum* Butler, Phytophthora blight caused by *Phytophthora drechsleri* Tucker f. sp. *cajani*, Macrophomina root rot caused by *Macrophomina phaseolina* (Tassi) Goid., stem canker caused by *Phoma cajani* (Rangel) and Alternaria blight caused by *Alternaria* sp. Fusarium wilt caused by *Fusarium udum* Butler, a soil and seed borne fungus spreads through wind, water and soil and can survive up to 3 years on infected plant debris and is of great economic importance (Shinde et al. 2014). Symptoms of the disease appear during flowering when the plant is just 1–2 months old. Likewise, *Phytophthora* blight another fungal disease caused by *Phytophthora drechsleri* Tucker f. sp. *cajani* is a common infection of *Cajanus cajan* (L.) Millsp. (Pande et al. 2011). It is a soilborne fungus and thus is fast spreading, surviving as dormant mycelia and chlamydospores in the soil. It is greatly affected by the weather. Rainy season favours the growth of the fungus. The spores of the fungus are spread through air and water. Warm and humid weather after the infection has occurred is a serious concern as it damages the plant and facilitates infection. Phoma stem canker of pigeon pea caused by *Phoma cajani* is one of the emerging diseases of the crop. The symptoms of the disease first appear on the stems as a necrotic spot and later turn into canker, resulting in the wilting of the whole plant. Macrophomina root rot is also among one of the important fungal infections of *Cajanus cajan* (L.) Millsp. caused by *Macrophomina phaseolina* (Tassi) Goid. This disease along with Alternaria blight caused by *Alternaria alternata* is a major problem for late-sown crops. Both these diseases are greatly affected by the weather. They are more prominent in hot and humid season. Under these conditions, root rot spreads to the base of the stem. The lesions further coalesce and cause the branches and then the entire plant to dry up and die.

12.3 Management of Disease

12.3.1 Cultural Management

Cultural practices are the traditional practices used by farmers to overcome diseases caused by pathogens in the crop. The commonly used practices include crop rotation, intercropping, interrow spacing, removal of diseased plant, spraying of nitrogen, etc. Verma and Rai in 2006 reported crop rotation with *Sorghum bicolour* (L.) Moench (sorghum), *Nicotiana tabacum* L. (tobacco) or *Ricinus communis* L. (castor) every 3 years terminates the pathogen from the field. They also stated that growing sorghum or fallow for 1 year on the same field of pigeon pea reduces the incidence of wilt disease up to below 20%. The spray of green manure with *Crotalaria juncea* reduces rot and wilt diseases to a great extent (Upadhyay and Rai 1981). The application of nitrogen as farmyard manure has also been found to be effective. One of the common and effective practices to control the diseases of pigeon pea is intercropping. Growing of other crops like sorghum or black gram as intercrop has proved to be effective (Table 12.1).

12.3.2 Chemical Management

Chemical management involves the treatment of the disease through chemical sprays. Numerous chemicals have been suggested for the management of fungal diseases of pigeon pea for long (Singh 1998). Pigeon pea seeds when treated with

Table 12.1 Cultural practices for disease control against some major fungal diseases

Disease	Common cultural practice
Fusarium wilt	• A field with no previous record (up to 3 years) of *Fusarium* wilt should be selected • Seeds used should be collected from disease-free fields of pigeon pea • The intercropping pattern is preferred • Rotation of 3 years and mixed cereal crops like sorghum, tobacco, etc. is beneficial • Solarization of soil in summer is also encouraged to reduce disease incidence
Phytophthora blight	• Field with no previous disease record is preferred • Sowing of seeds should be avoided in waterlogging areas like the low-lying patch • Good drainage should be ensured through raised seedbeds • Interrow spacing also proves to be helpful
Dry root rot	• Field with no previous disease record is preferred • Late sowing of seeds should be avoided to reduce the risk of high temperature and drought conditions
Phoma stem canker	• Field with no previous disease record is preferred • Infected plants should be removed subsequently to reduce the spread of infection
Alternaria blight	• Seeds used for sowing should be taken from healthy fields • Avoid late sowing of the crop

Table 12.2 Chemical practices for disease control fungal diseases

Disease	Chemical practice
Fusarium wilt	• Seed bacterization with Benlate and thiram in 1:1 (3 g per kg of seed)
Phytophthora blight	• Foliar spray at 15 days interval with Ridomil MZ (2 sprays)
Dry root rot	• Dressing of seeds with tolclofosmethyl or thiram
Alternaria blight	• Foliar spray with Indofil M45

an equal part mixture of benomyl and thiram eradicate the disease (ICRISAT 1987; Reddy et al. 1993). Supplementing soil with boron, manganese or zinc and methyl bromide (CH_3Br) reduces the incidence of fusarium wilt. Ingole et al. (2005) also reported similar findings with a mixture of carbendazim + thiophanate (0.15 + 0.10%) against wilt disease of pigeon pea. Few antibiotics like bulbiformin have also found to be an effective tool against pathogens (Table 12.2).

12.3.3 Biological Management

The application of hazardous fungicides affects the environment in adverse ways, and moreover, chemical fertilizers are not targeted specifically. It not only degrades the ecosystem but also has negative effects on human health. Fungicides affect the food chain as they are toxic to species like earthworms and microorganisms and also to an extent affect genotoxicity of humans (Shuping and Eloff 2017). They cause water and soil pollution too. The solution to this above problem lies in sustainable agriculture. The application of potential microorganisms which are part of the existing ecosystem serves as an effective means against plant protection system. Biological management of diseases has been reported by several workers and serves as an attractive tool for eco-friendly management of soilborne as well as other pathogens degrading the crop. Disease incidence of fusarium wilt has been reduced by the application of antagonistic microorganisms like fungi and bacteria (Passari et al. 2017; Anjaiah et al. 2003; Mandhare and Suryawanshi 2005; Maisuria et al. 2008; Singh et al. 2002). Out of cluster of scientific reports, few of them have notable biological measures that are functional for the management of pigeon pea diseases. Seed inoculation with rhizosphere bacteria like *Bacillus subtilis*, *Pseudomonas fluorescens and Pseudomonas aeruginosa* is very effective against fungal disease of pigeon pea (Mahesh et al. 2010). Integrated management strategies (IDM) which involve a combination of fungicides and biocontrol agents also prove to be beneficial for the management of *Fusarium udum* Butler (Pande et al. 2012). Oil formulations of *Trichoderma* strains like *Trichoderma harzianum* reduce the traces of soilborne pathogens from the diseased plants (Khan and Khan 2002). Siddiqui and Shakeel (2007) suggested that various rhizobacteria are efficient biocontrol agents. Plant extracts like neem and eucalyptus, garlic and henna, ginger and tulsi are also found to have an inhibitory effect against Alternaria blight of pigeon pea (Rathore et al. 2018).

12.4 Biocontrol Agents

The property of microorganisms to fight against phytopathogens is termed as a form of biological control (Duffy and Defago 2009). This approach is eco-friendly, much effective as well as cost-efficient. These PGPRs produce antifungal metabolites, creating competition for nutrients that act as chief modes of biocontrol activity (Duffy and Defago 2009). Rhizobacteria produce some antifungal metabolites like HCN, phenazines, pyoluteorin and tensin which kill the fungal pathogen (Bhattacharyya and Jha 2012). *Bacillus* spp. (Gong et al. 2006) and *Pseudomonas* (Leonardo et al. 2006) are two PGPRs that have been reported being effective biocontrol agents. Among these bacterial species, *Bacillus subtilis*, *Bacillus amyloliquefaciens* and *Bacillus cereus* are the most effective ones for controlling plant diseases through various mechanisms (Passari et al. 2016a; Francis et al. 2010). PGPRs like *Bacillus* spp. and *Pseudomonas* spp. have this ability to make endospores which allows them to sustain in a wide range of environmental conditions and hence make them efficient biofertilizers (Perez-Garcia et al. 2011). Application of *T. harzianum*, *T. viride*, *B. subtilis* and *P. fluorescens* when mixed with neem or karanj cake and compost not only reduces the diseases but also enhances the longevity of biocontrol agents (Narayanan et al. 2015; Shanmugapackiam et al. 2016).

Application of biocontrol agents can be done in three forms:

1. By application of fungi
2. By application of AMF
3. By application of bacteria

12.4.1 By Application of Fungi

Trichoderma sp. secretes secondary metabolites which are antifungal and hence has great potential to act as biocontrol agents. They reduce the fungal pathogen either directly by mycoparasitism or through indirect mechanisms like competition for nutrients and space to survive and modifications of environmental conditions. They help in the promotion of plant growth and also activate the defence mechanism of the plant. Whipps and Lumsden (2001) stated that species of *Trichoderma* have been widely accepted as biocontrol agents against numerous phytopathogens. *Trichoderma* species are useful virulent saprophytes that act as biocontrol agents against phytopathogenic fungi by various mechanisms such as rhizosphere competition, mycoparasitism and antibiotic and enzyme production and induce resistance. Growth promotion activity of *Trichoderma* has also been reported (Cumagun 2012; Harman et al. 2004). Strains of *Trichoderma* (*T. viride*, *T. harzianum*, *T. virens*) were evaluated under field conditions against *Fusarium udum*; out of which *T. viride* was found to be most promising at 15% concentration (Chaudhary et al. 2017). The inoculation of seeds with antagonists helps in externally managing seed and soilborne pathogens. Talc-based formulation of *Trichoderma* sp. has been used to coat seeds.

12.4.2 By Application of AMF

AMF or arbuscular mycorrhizal fungi are the groups of fungi that act as promising biofertilizers. Dumas-Gaudot et al. (2000), Garmendia et al. (2005) and Garcia-Garrido (2009), in their respective studies, reported that AMF-mediated bioprotection is accepted as a key practice for disease control. AMF is currently exploited for its anti-pathogenic properties. Linderman (2000) reported that induced systematic resistance or ISF is the mechanism behind AMF phytoprotection. This mechanism concentrates more on nutritional changes like competition with infection sites, changes in the morphology of root and shoot tissues, abiotic stress reduction and changes in the mycorrhizosphere and chemicals, constituting changes in plant tissues (Hause and Fester 2005). All these properties make AMF a good biofertilizer also in the coming future.

12.4.3 By Application of Bacteria

Plant growth-promoting bacteria are the bacteria present in rhizospheric soil which enhance the growth of the plant directly or indirectly. The awareness of PGPR is increasing steadily in the world. They are applied to several economically important crops to increase the yield of the crop by enhancing the growth of the plant and protecting it from different pathogens. PGPR promotes plant growth by procurement of minerals like phosphorous, nitrogen, etc. directly from the soil (Gyaneshwar et al. 1998) and also indirectly by acting against plant pathogens as a biocontrol agent. Several reports suggest an increment in the quality and the number of different crops worldwide through the application of PGPRs under normal as well as stressed conditions (Passari et al. 2019). The application of PGPR is encouraged because it reduces the dependence on hazardous chemical fertilizers for improving plant growth and helps in reducing plant pathogens, which destabilizes the agriculture system. PGPR exhibits positive effect on the germination of the seeds, the yield of the crop and their tolerance towards stresses like drought and salt (Passari et al. 2019; Brown 1974). PGPR is an effective antagonist against plant pathogens like *Fusarium udum* and *Macrophomina phaseolina*. Soil microbe's interaction with the rhizosphere plays an important role in solubilizing and mobilizing a limited amount of nutrients available and also their uptake by the plant (Bolton et al. 1993; Mantelin and Touraine 2004). PGPR has beneficial effects as a biocontrol agent to important crops like legumes, cereals, fruits, vegetables, etc. According to reports, the exact estimate is unknown, but an average of more than 50% of crop losses in pigeon pea is due to pathogenic microorganisms (Rajash 2005). Thus, the need of the hour is to exploit and enhance the efficacy of soilborne control agents and use their best possible combination against plant pathogens (Mishra et al. 2016; Chang et al. 2005). The encouragement for the use of PGPR as biofertilizers against plant pathogens will serve as a promising alternative to deadly chemical fertilizers and pesticides (Goldstein 1995). Screening of soil for bacterial antagonist against pathogens is a notable biological advancement (Passari et al. 2016a; Karimi et al. 2012; Siddiqui et al. 2005), mostly for PGPR as a biocontrol agent (Siddiqui and Shakeel 2007; Prasad et al. 2002). Inoculation of *Pseudomonas aeruginosa* in the seed is effective against fusarium wilt disease of pigeon pea (Mahesh et al. 2010).

12.4.3.1 Modes of Action of PGPR

The mechanism of action of PGPR is not completely known; however, they are reported to exhibit several beneficial activities for plant growth promotion (Khan et al. 2009; Zaidi et al. 2009). PGPR promotes plant growth in two ways: directly and indirectly (Glick 2012). Pigeon pea is the most staple and proteinaceous food available in many developing countries; hence, it becomes important to protect this crop from damage. Root-nodulating bacteria *Sinorhizobium* inhibited the growth of fusarium wilt of pigeon pea as it possesses chitinase and β-gluconase production (Kumar et al. 2010). Plant growth promotion takes place indirectly when PGPR increases plant growth by decreasing the activity of plant pathogens (Xiang et al. 2017).

12.4.3.1.1 Nitrogen Fixation

Nitrogen is a vital nutrient required for the growth and productivity of the plant. The atmospheric N_2 is converted into plant-utilizable forms by biological N_2 fixation during which nitrogen gets converted into ammonia, and this is done with the help of nitrogen fixation bacteria present in the rhizospheric soil catalysed by nitrogenase enzyme (Kim and Rees 1994). Biological nitrogen fixation, also known as BNF, usually takes place at mild temperatures, by widely spread nitrogen-fixing bacteria (Raymond et al. 2004). This provides an economically beneficial and environmentally friendly alternative to chemical fertilizers (Ladha et al. 1997). Nitrogen-fixing bacteria (symbiotic bacteria) show symbiosis with plants belonging to leguminosae family like rhizobia (Ahemad and Khan 2011; Zahran 2001) However, non-symbiotic nitrogen-fixing bacteria provide only a small amount of the fixed nitrogen that bacterially associated host plant requires (Glick 2012).

12.4.3.1.2 Phosphate Solubilization

After nitrogen, phosphorus is the second most vital nutrient required for plant growth. This is also abundantly available both in an organic and inorganic form in the soil (Khan et al. 2009). The low availability of phosphorous to the plants is due to its presence in the insoluble form which plants are not able to absorb (Bhattacharyya and Jha 2012). The only soluble form of phosphorous available for the use of plants is monobasic and dibasic (Jha and Saraf 2015). To fulfil the phosphorous requirement, phosphatic fertilizers are given as a supplement in the fields. As plants do not absorb the full amount of applied fertilizer, the rest gets converted into insoluble complexes in the soil (McKenzie and Roberts 1990). This practice not only affects the environment but is also not cost-effective. Hence finding a better reliable solution to this problem is necessary. PGPR has coupled with phosphate solubilizing activity which may provide the available phosphorous to the plants in a much eco-friendly way (Khan et al. 2006).

12.4.3.1.3 Siderophore Production

Iron is a prominent nutrient available for all lives possible on earth. It is needed by all living beings.

In properly aerated soils, iron in the form Fe^{3+} (ferric iron), which is easily precipitated as iron oxide, is absorbed by plants (Duffy 1994). This property of microbes to secrete siderophores makes them suitable biocontrol agents as they induce competition for iron availability in the rhizosphere, hence restricting the proliferation of fungal phytopathogens in the vicinity of the crop, because of less availability of iron. CAS or chrome azurol agar media is used to isolate siderophore-producing bacteria. Rajkumar et al. (2008) have reported the growth of the plant through siderophore, because of the siderophore-producing bacteria in the rhizosphere.

12.4.3.1.4 Phytohormone Production

Microbes are known to synthesise phytohormones like auxins or IAA, i.e. indole acetic acid, for a long time. About 80% of the microbes isolated from the rhizosphere, of many crops, secrete secondary metabolites like auxins (Patten and Glick 1996). Indole acetic acid has a prominent function in bacteria-plant interactions (Passari et al. 2016a, b; Spaepen and Vanderleyden 2011). It is also reported that IAA has a plant defence mechanism against plant pathogens, and it produces a signalling effect to reduce the IAA production by the plant pathogen (Spaepen and Vanderleyden 2011).

12.5 Microbial Consortium

Most applications of biocontrol of plant diseases use single biocontrol agents as the antagonist against plant pathogens. The microbial consortium works well as, biopesticides, against a wide spectrum of plant pathogens which is a little difficult to be fulfilled using a single biocontrol agent. Biocontrol agents individually or in consortium attack pathogens through antagonism effect. They act better and more effectively when combined and when belonging to the same ecosystem. Vital and future promising candidates of the microbial consortium are *Trichoderma* sp., *Pseudomonas* sp. and *Bacillus* sp. Seed bacterization with a consortium of *Rhizobium* and *Pseudomonas putida*, *P. fluorescens* and *Bacillus* increased yield and biomass of pigeon pea crop (Tilak et al. 2006). *Trichoderma* sp. in association with AMF has great potential against plant pathogens (Wehner et al. 2010). The consortium of bio-organic (municipal waste) and applied organic (*Rhizobium* sp.) showed prominent improvement in the growth of pigeon pea over control plant (Rizwan and Mahmood 2017). Didwania et al. (2019) have also reported integrated management for Alternaria blight in oil-yielding crops.

12.6 Biotechnological Approaches to Biological Management

The detailed information on biotechnological techniques and genetics is important for developing a mechanism against susceptible varieties. Numerous resistant theories are known against fusarium wilt, and hence a single dominant gene has been established (Owuoche and Silim 2010; Kotresh et al. 2006). Many well-characterized or little-known genes, earlier reported being involved in legume crops, defend against fungal infection in pigeon pea. Resistant varieties available in the market against Phytophthora blight are Hy 4, ICPL 150, ICPL 288, ICPL 304, KPBR 80-1-4 and KPBR 80-2-1 (ICAR database). Out of 80 entries evaluated under sick plot, 18 entries WRP-1, BDN-2004-1, MAHABEJ, BRG-14-2, PT-257, BRG-14-1, MA-13, BWR-133, GRG-160, IPA8F, KA-12-03, ICPL-87119, KPL-44, KPL-43, BSMR571, BSMR-846, BSMR-579 and BSMR-2 have showed moderate resistant reaction with 0.00–10.00 per cent disease incidence. Similarly, Mishra and Dhar (2005) reported the same findings in vitro. Prasanthi et al. (2009) have reported a disease score of zero in treated and untreated pots of genotype ICP 8863, in pot culture screening technique against fusarium wilt-resistant/fusarium wilt-susceptible genotypes. IVT-520, IVT-509 and AVT-603 were found to be resistant against pod bug damage among 29 genotypes screened (Singh et al. 2017).

12.7 Conclusion

With the increasing population of the world, the demand for staple food like legumes, which are rich in protein, would also increase. Hence measures are required to fulfil the demand of the crop.

Decades ago the green revolution happened which increased the agriculture supply globally. This revolution saved the then population from hunger and malnutrition but, in turn, also triggered the use of chemical fertilizer. These chemical fertilizers are very harmful to our environment as they enter the food chain. So it is the need of the hour that we adapt better means to improve the quality as well as quantity of the crop but keeping in mind the environment safety also. Biofertilizers are an excellent solution to this problem of chemical fertilizers. Biofertilizers help in the improvement of plant growth and also act as biocontrol agents. They are eco-friendly and cost-effective means for crop improvement. Their use will serve as an instrument to ensure productivity and stability which will lead us to perfect agricultural practices in the world. A combination of biotechnological approaches with microbial consortium can contribute to go a long way in fighting with fungal diseases of pigeon pea and also to increase the yield.

References

Ahemad M, Khan MS. Effects of pesticides on plant growth promoting traits of *Mesorhizobium* strain MRC4. Bull Environ Contam Toxicol. 2011;86(4):384–8.

Ahlawat IPS, Gangaiah B, Singh IP. Pigeonpea (*Cajanus cajan*) research in India—an overview. Indian J Agri Sci. 2005;75:309–20.

Anjaiah V, Cornelis P, Koedam N. Effect of genotype and root colonization in biological control of Fusarium wilts in pigeonpea and chickpea by *Pseudomonas aeruginosa* PNA1. Can J Microbiol. 2003;49(2):85–91.

Bhattacharyya PN, Jha DK. Plant growth-promoting rhizobacteria (PGPR): emergence in agriculture. World J Microbiol Biotech. 2012;4:1327–50.

Brown ME. Seed and root bacteization. Annu Rev Phytopatol. 1974;12:181–97.

Chang C, Wang YF, Kanamori Y, Shih JJ, Kawai Y, Lee CK, Wu KC, Esashi M. Etching submicrometer trenches by using the Bosch process and its application to the fabrication of antireflection structures. Journal of micromechanics and microengineering. 2005;15(3):580.

Chaudhary B, Kumar S, Kushwaha S. Bio-efficacy of *Trichoderma* species against Pigeon pea wilt pathogen. JANS. 2017;9(2):1–6.

Cumagun CJR. Managing plant diseases and promoting sustainability and productivity with Trichoderma: the Philippine experience. J Agri Sci Tech. 2012;14:699–714.

Didwania N, Gupta KN, Gupta S, Bisen R. Bio-intensive approaches in the management of fungal diseases of oil yielding crops in Bio-intensive approaches: Application and effectiveness in plan diseases management. Today & Tomorrow's Printers and Publishers, New Delhi – 110 002, India 333–356; 2019.

Duffy BK, Defago G. Environmental factors modulating antibiotic and siderophore biosynthesis by *Pseudomonas fluorescens* biocontrol strains. Appl Environ Microbiol. 2009;65(6):2429–38.

Dumas-Gaudot E, Gollotte A, Cordier C, Gianinazzil S, Gianinazzi-Pearson V. Modulation of host defence systems. In: Kapulnik Y, Douds DD, editors. Arbuscular mycorrhizas: physiology and function. Dordrecht: Kluwer; 2000. p. 173–200.

FAO. Food and Agriculture Organization of the United Nations, Rome, Italy "FAOSTAT" www.fao.org; 2018.

Francis I, Holsters M, Vereecke D. The gram-positive side of plant-microbe interaction. Environ Microbial. 2010;12:1–12.

Garcia-Garrido JM. Arbuscular mycorrhizae as defense against pathogens. In: White JF, Torres MS, editors. Defensive mutualism in microbial symbiosis. Boca Raton, FL: CRC Press; 2009. p. 183–98.

Garmendia I, Goicoechea N, Aguirreolea J. Moderate drought influences the effect of arbuscular mycorrhizal fungi as biocontrol agents against Verticillium-induced wilt in pepper. Mycorrhiza. 2005;15:345–56.

Glick BR. Plant growth-promoting bacteria: mechanisms and applications. *Scientifica*. 2012.

Goldstein AH. Evidence for mutualism between a plant growing in a phosphate-limited desert environment and a mineral phosphate solubilizing rhizobacteria. FEMS Microbiol Ecol. 1995;30(4):295–300.

Gong M, Wang JD, Zhang J, Yang H. Study of the antifungal ability of *Bacillus subtilis* strain PY-1 in vitro and identification of its antifungal substance (IturinA). Acta Biochim Biophys Sin. 2006;38(4):233–40.
Gyaneshwar P, Naresh KG, Parekh LJ. Effect of buffering on the P-solubilizing ability of microorganisms. World J Microbial Biotechnol. 1998;14(5):669–73.
Harman GE, Howell CR, Viterbo A, Chet I, Lorito M. *Trichoderma* species: opportunistic, a virulent plant symbionts. Nat Rev Microbiol. 2004;2:43–56.
Hause B, Fester T. Molecular and cell biology of arbuscular mycorrhizal symbiosis. Planta. 2005;221:184–96.
ICRISAT Annual Report (Patancheru, Andhra Pradesh, India: International Crops Research Institute for the Semi-Arid Tropics) 1987;177.
Ingole MN, Ghawade RS, Raut BT, Shinde VB. Management of Pigeonpea wilt caused by *Fusarium udum Butler*. Crop Prot Product. 2005;1(2):67–9.
Jha CK, Saraf M. Plant growth promoting rhizobacteria (PGPR): a review. J Agric Res Dev. 2015;5:108–19.
Karimi R, Owuoche JO, Silim SN. Importance and management of Fusarium wilt (*Fusarium udum* Butler) of pigeonpea. Intl J Agron Agri Res. 2012;2:1–14.
Khan A, Williams KL, Nevalainen HK. Control of plant-parasitic nematodes by Paecilomyces lilacinus and Monacrosporium lysipagum in pot trials. Biocontrol. 2006;51(5):643–658.
Khan MR, Khan SM. Effects of root-dip treatment with certain phosphate solubilizing microorganisms on the Fusarial wilt of tomato. Bioresour Technol. 2002;85(2):213–5.
Khan MS, Zaidi A, Wani PA, Oves M. Role of plant growth promoting rhizobacteria in the remediation of metal contaminated soils. Environ Chem Lett. 2009;7(1):1–19.
Kim J, Rees DC. Nitrogenase and biological nitrogen fixation. Biochemist. 1994;33:389–97.
Kotresh H, Fakrudin B, Punnuri SM, Rajkumar BK, Thudi M, et al. Identification of two RAPD markers genetically linked to a recessive allele of a Fusarium wilt resistance gene in pigeonpea (*Cajanus cajan L.* Millsp.). Euphytica. 2006;149(1–2):113–20.
Kumar H, Bajpai VK, Dubey RC. Wilt disease management and enhancement of growth and yield of *Cajanus cajan* (L) var. Manak by bacterial combinations amended with chemical fertilizer. Crop Protect. 2010;29:591–8.
Ladha JK, Bruijn FJ, Malik KA. Introduction: assessing opportunities for nitrogen fixation in rice-a frontier project. Plant Soil. 1997;124:1–10.
Leonardo D, Blanca LF, Landa B, Weller DM. Host crop affects rhizosphere colonization and competitiveness of 2,4-diacetylphloroglucinol-producing *Pseudomonas fluorescens*. Phytopathology. 2006;96:751–62.
Linderman RG. Effects of mycorrhizas on plant tolerances to diseases. In: Kapulnik Y, Douds DD, editors. Arbuscular mycorrhizas: physiology and function. Dordrecht: Kluwer; 2000. p. 345–65.
Mahesh M, Saifulla M, Sreenivasa S, Shashidhar KR. Integrated management of pigeonpea wilt caused by *Fusarium udum Butler*. J Biol Sci. 2010;2:1–7.
Maisuria VB, Gohel V, Mehta AN, Patel RR, Chhatpar HS. Biological control of Fusarium wilt of pigeon pea by *Pantoea dispersa*, a field assessment. Ann Microbiol. 2008;58(3):411–9.
Mandhare VK, Suryawanshi AV. Application of Trichoderma species against pigeonpea wilt. JNKVV Res J. 2005;32(2):99–100.
Mantelin S, Touraine B. Plant growth-promoting bacteria and nitrate availability: impacts on root development and nitrate uptake. Journal of experimental Botany. 2004;55(394):27–34.
McKenzie RH, Roberts TL. Soil and fertilizers phosphorus update. In: Proceedings of Alberta Soil Science Workshop Proceedings Feb. 20–22 Edmonton Alberta; 1990. pp. 84–104.
Mishra S, Dhar V. Efficient method of inoculation by *Fusarium udum,* the incident of pigeon pea wilt. Indian Phytopath. 2005;58(3):332–4.
Mishra VK, Passari AK, Singh BP. In vitro antimycotic and biosynthetic potential of fungal endophytes associated with *Schima wallichii*. In: Kumar P, et al., editors. Current trends in disease diagnostics: Springer; 2016. p. 367–81.

Narayanan P, Vanitha S, Rajalakshmi J, Parthasarathy S, Arun K, et al. Efficacy of biocontrol agents and fungicides in management of mulberry wilt caused by *Fusarium solani*. J Biol Control. 2015;29(2):107–14.

Owuoche RKJO, Silim SN. Inheritance of Fusarium wilt resistance in pigeon pea *Cajanus cajan* (L.) Mill sp. Ind J Genet Plant Br. 2010;70(3):271–6.

Pande S, Sharma M, Mangla UN, Ghosh R, Sundaresan G. Phytophthora blight of Pigeonpea [*Cajanus cajan* (L.) Millsp.]: an updating review of biology, pathogenicity and disease management. Crop Prot. 2011;30(8):951–7.

Pande S, Sharma M, Avuthu N, Telangre R. High throughput phenotyping of chickpea diseases: stepwise identification of host plant resistance. Information Bulletin No. 2012; 92. ICRISAT.

Passari AK, Mishra VK, Leo VV, Gupta VK, Singh BP. Phytohormone production endowed with antagonistic potential and plant growth promoting abilities of culturable endophytic bacteria isolated from *Clerodendrum colebrookianum* Walp. Microbiol Res. 2016a;193:57–73.

Passari AK, Chandra P, Zothanpuia, Mishra VK, Leo VV, et al. Detection of biosynthetic gene and phytohormone production by endophytic actinobacteria associated with *Solanum lycopersicum* and their plant-growth-promoting effect. Res Microbil. 2016b;167:692–705.

Passari AK, Lalsiamthari PC, Zothanpuia, Leo VV, Mishra VK, et al. Biocontrol of Fusarium wilt of *Capsicum annuum* by rhizospheric bacteria isolated from turmeric endowed with plant growth promotion and disease suppression potential. Eur J Plant Pathol. 2017;150:831–46.

Passari AK, Upadhyaya K, Singh G, Abdel-Azeem AM, Thankappan S, et al. Enhancement of disease resistance, growth potential, and photosynthesis in tomato (*Solanum lycopersicum*) by inoculation with an endophytic actinobacterium, *Streptomyces thermocarboxydus* strain BPSAC147. PLoS One. 2019;14(7):e0219014.

Patten CL, Glick BR. Bacterial biosynthesis of indole-3-acetic acid. Canadian journal of microbiology. 1996;42(3):207–220.

Perez-Garcia A, Romero D, Vicente D. Plant protection and growth simulation by microorganism: biotechnological applications of Bacillus in agriculture. Curr Open Biotech. 2011;22:187–93.

Prasad RD, Rangeshwaran R, Hegde SV, Anuroop CP. Effect of soil and seed application of *Trichoderma harzianum* on pigeon pea wilt caused by *Fusarium udum* under field conditions. Crop Prot. 2002;21(4):293–7.

Prasad P, Doharey RK, Singh SN, Singh RK, Kumar M, et al. Communication and psychological behaviour of the pigeon pea growers in Chitrakoot district, India. Int J Curr Microbiol App Sci. 2017;6:2032–7.

Prasanthi L, Reddy BBV, Rani RK, Naidu HP. Molecular markers for screening Fusarium wilt resistance in pigeon pea (*Cajanus cajan L. Millsp*). Legume Res. 2009;32(1):19–24.

Rajash P. Effect of plant growth promoting rhizobacteria on canola (*Brassica napus* L) and lentil (*Lens culinaris*) plant ETD project; 2005.

Rajkumar M, Ma Y, Freitas H. Characterization of metal-resistant plant-growth promoting *Bacillus weihenstephanensis* isolated from serpentine soil in Portugal. J Basic Microbiol. 2008;48:500–8.

Rathore US, Singh SK, Kumar S, Saloni R. Application of botanicals for effective management of Alternaria blight of Pigeon pea. J Pharmacog Phytochem. 2018;2:328–38.

Raymond J, Siefert JL, Staples CR, Blankenship RE. The natural history of nitrogen fixation. Mol Biol Evol. 2004;2:541–54.

Reddy MV, Raju TN, Sharma SB, Nene YL, McDonald D. Handbook of pigeon pea diseases, (In En. Summaries in En. Fr.). Information Bulletin. 14. Patancheru, A.P. 502 324 India: International Crops Research Institute for the Semi-Arid Tropics 64; 1993.

Rizwan AA, Mahmood I. Optimization of organic and bio organic fertilizers on soil properties and growth of pigeon pea. Scientia Horticult. 2017;226:1–9.

Saxena KB, Kumar RV, Sultana R. Quality nutrition through pigeon pea—a review. Health. 2010;2(11):1–10.

Shanmugapackiam SS, Parthasarathy M, Jebaraj D, Christopher DJ. Exploitation of compost amended biological agents for the management of vascular wilt *Fusarium oxysporum* f. sp. *lycopersici* in tomato. Adv Life Sci. 2016;5(7):2933–41.

Shinde VS, Zagade SN, Chavan A. A Cultural and morphological variation in *Fusarium udum*. J Plant Dis Sci. 2014;9(2):237–44.
Shuping DSS, Eloff JN. The use of plants to protect plants and food against fungal pathogens: a review. Afr J Tradit Complement Altern Med. 2017;14(4):120–7.
Siddiqui ZA, Shakeel U. Screening of Bacillus isolates for potential biocontrol of the wilt disease complex of pigeon pea (*Cajanus cajan*) under greenhouse and small-scale field conditions. J Plant Pathol. 2007;89(2):179–83.
Siddiqui S, Siddiqui ZA, Iqbal A. Evaluation of fluorescent pseudomonas and Bacillus isolates for the biocontrol of wilt disease complex of pigeon pea. World J Microbiol Biotech. 2005;21:729–32.
Singh R. Chemical control of Fusarium wilt of pigeon pea. Kor J Mycol. 1998;26(4):416–23.
Singh R, Singh BK, Upadhyaya SK, Rai B, Lee YS. Biological control of Fusarium wilt disease of pigeonpea. J Plant Pathol. 2002;18(5):279–83.
Singh I, Shankar U, Abrol DP, Mondal A. Diversity of insect pollinators associated with pigeon pea, *Cajanus cajan* L. Mill sp. and their impact on crop production. Int J Curr Microbiol App Sci. 2017;6(9):528–35.
Spaepen S, Vanderleyden J. Auxin and plant-microbe interactions. Cold Spring Harbor perspectives in biology. 2011;3(4), p.a001438.
Tilak KVB, Ranganayaki N, Manoharachari C. Synergistic effects of plant-growth promoting rhizobacteria and rhizobium on nodulation and nitrogen fixation by Pigeonpea (Cajanus cajan). Eur Soil Sci. 2006;57(1):67–71.
United Nations Public Administration Network. 2010. http://www.un.org/apps/news/story.asp?NewsID #.VtkHC7alzDc.
Upadhyay RS, Rai B. Effect of cultural practices and soil treatments on incidence of wilt disease of pigeonpea. Plant Soil. 1981;62:309–12.
Wehner JP, Antunes PMM, Powell JR, Mazukatow J, Rillig MC. Plant pathogen protection by arbuscular mycorrhizas: a role for fungal diversity. Pedobiologia. 2010;53:197–201.
Whipps JM, Lumsden RD. Commercial use of fungi as plant disease biological control agents: status and prospects. In: Butt T, Jackson C, Magan N, editors. Fungal biocontrol agents: progress, problems and potential 9–22. Wallingford: CABI Publishing; 2001.
Xiang N, Lawrence KS, Kloepper JW. Biological control of *Meloidogyne incognita* by spore-forming plant growth-promoting rhizobacteria on cotton. Plant Dis. 2017;101(5):774–84.
Zahran HH. Rhizobia from wild legumes: diversity, taxonomy, ecology, nitrogen fixation and biotechnology. J Biotech. 2001;91(2–3):143–53.
Zaidi A, Khan MS, Ahemad M, Oves M. Plant growth promotion by phosphate solubilizing bacteria. Acta Microbiol Immunol Hung. 2009;56(33):263–84.

Index

B. P. Singh et al. (eds.), *Management of Fungal Pathogens in Pulses*, Fungal Biology, https://doi.org/10.1007/978-3-030-35947-8

The manufacturer's authorised representative in the EU is Springer Nature Customer Service Centre GmbH, Europaplatz 3, 69115 Heidelberg, Germany. If you have any concerns regarding our products, please contact ProductSafety@springernature.com

Printed and bound by CPI Group (UK) Ltd, Croydon, CR0 4YY
30/09/2026
02225458-0006